W9-DEK-381

A History
of
Delusions

A History
of
Delusions

....

*The Glass King,
a Substitute Husband
and a Walking Corpse*

....

VICTORIA SHEPHERD

A Oneworld Book

First published by Oneworld Publications in 2022

Copyright © Victoria Shepherd 2022

ISBN 978-0-86154-091-4
eISBN 978-0-86154-092-1

Typeset by Tetragon, London
Printed and bound in Great Britain by Clays Ltd, Elcograf S.p.A.

Illustration credits: Postcards of Atelier Stockman, Atelier 28 Henrion and Le secours de guerre from author's collection. Map of faux Paris © *Illustrated London News*/Mary Evans Picture Library. Air Loom from *Illustrations of Madness* courtesy of the Wellcome Collection. Incurable patient register © Bethlem Museum of the Mind. James Tilly baptism record reproduced courtesy of Staffordshire Record Office. Marriage bond of James Tilly Matthews © London Metropolitan Archives, City of London. Mesmer's baquet © Archivah/Alamy Stock Photo. Robert Burton portrait reprinted with kind permission of the Principal and Scholars of the King's Hall and College of Brasenose in Oxford. Robert Burton bust reprinted by kind permission of the Dean & Canons of Christ Church. Pieter Codde, *Young Scholar in His Study, Melancholy* © Maidun Collection/Alamy Stock Photo. Simon Forman casebook © Bodleian Library. Portrait of Henry Percy courtesy of Rijksmuseum. *Ars Moriendi* woodcuts from archive.org. King Charles from Froissart's *Chronicles* courtesy of Wikimedia Commons. Margaret Nicholson portrait © Trustees of the British Museum. Bedlam admission register © Bethlem Museum of the Mind. *Louis Capet Being Welcomed to Hades* licensed under CC0 from Musée Carnavalet. Napoleon portrait © Bridgeman Images. Eugène Atget photograph courtesy of Wikimedia Commons. De Clérambault drapes © RMN-Grand Palais/Dist. Photo SCALA, Florence.

Oneworld Publications
10 Bloomsbury Street
London WC1B 3SR
England

To Mohit and Kit,
and to my parents, Ann and Jack

Et qui n'est chaque fois ni tout a fait la même,
ni tout a fait une autre.

'And each time she is neither entirely the same,
nor entirely different.'

'Mon rêve familier'
PAUL VERLAINE, 1866

In delusions, everything which one wishes and fears may
find its level of expression and as far as can be judged by the
present state of our knowledge, many other things, perhaps
even everything which can be experienced or thought.

EUGEN BLEULER, 1911

Contents

Preface

The 1340s, Rouen, northern France, behind a wonky timbered building near the river Seine, at the back of the rear courtyard, a furnace glows fierce orange. A gaffer stokes; he needs it as hot as it can get. He is working on something completely new. It will make his family rich. He is having one last go and he wipes his sweaty brow across the crook of his arm, then shovels out another measure of sand from the bucket into the crucible, and then a measure of ash. He mixes the batch together and pushes it into the roaring furnace. Now he waits for the mixture to heat. The batch turns pale. The temperature's high enough for the reaction. He takes his blowpipe to the furnace. Then he begins gathering up the molten substance like toffee, layer upon layer, until he has an orange globe on the end of the pipe. He puts his mouth to the pipe and blows hard. The globe inflates. He thrusts it back in the fire and starts to spin, watching as centrifugal forces push it out into a flat disk bullion. Now he carefully detaches the bullion from the blowpipe. In the middle of the disc, a 'bullseye' marks the point where the pipe joined the gather. Around the bullseye the material is thinner and can be cut into diamonds for use in windows. That's the stuff he's interested in. It looks good so far, but he's had his hopes up before. He watches it cool. The sand begins to lose its crystalline structure and gain an entirely new one – on a molecular level somewhere between a liquid and a solid. It will be hard like the more primitive version, but, if he can perfect it, there will be a key difference... He steadies his hand and lifts it up to inspect. There's the workshops' crooked roof-line clean against the sky; the workbench; his shoes. No need to squint, there's no milky blur. It's clear. It will bring a breathtaking new sharpness to windows. People may even want it to magnify objects. The gentry will love it. They will flock to Rouen and buy as much of it as he can make. He pauses. It's struck him that this will change more than the weight of his purse. It will transform the way people see the world.

The introduction of this innovative product, manufactured from what seemed an almost alchemical process – sand trans-formed through fire into something fragile and transparent – did have a profound impact on some parts of French society. The wealthy and noble classes enjoyed this new 'crown' glass in their own homes and they carried it carefully, tapped it gingerly, looked through it with wonder, saw what happened if dropped.

There's a twist. The effect was stranger than the Rouen glass-maker could possibly have imagined. Crown glass did indeed alter the way people saw the world, but it wasn't by virtue of the clarity of the view through the new windows. The transformative power of this material, its alchemy, was continuing to work on the people who had brought it into their homes. It was infiltrating them, influencing them at a deeper level. It was changing how they saw *themselves*. A few started to believe the chemical reaction was at work within their own body. Something was happening to their legs, their arms, their feet... They were turning into glass. Bits of them were now made out of it; translucent, brittle, fragile. Here was a startling example of how external processes might affect *inner* processes and create a delusion to moderate a person's relationship with the world.

A fifteenth-century French king, Charles VI, made the phenomenon famous. He underwent his own glassy metamorphosis in front of alarmed courtiers. Pope Pius II recorded in his chronicles that Charles had iron rods sewn into his clothes to prevent his glass bones breaking if he touched someone, and he is reported to have wrapped himself in blankets to protect against the danger of shattered buttocks.[1] We can picture him locked in the attrition of the Hundred Years War, yet privately consumed with anxiety about any hard surfaces which might come into contact with his rear end and frantically sourcing prophylactic soft furnishings. News of his belief leaked out and he offered the courts of Europe a good laugh, but it was nervous laughter. He had set off a chain reaction of Glass Men across the continent.

'Glass delusion', as the condition became known, is just one of the strange and compelling psychological phenomena that the history of delusions offers up to us.

The content and context of delusions change, era to era, person to person, over the centuries but common features remain. Delusions carry painfully insistent demands, and, for the person experiencing the delusion, the stakes are invariably high. They are often life and death. Charles VI orders his associates to back off – he will smash into pieces if they touch him. It's an absurd premise, but beneath the absurdity the perceived jeopardy is painfully real. If you pay closer attention to each of these historical accounts, you can pick up a series of urgent communiqués. Each story then takes on a quality of a psychological thriller for the audience. What does this person need us to know? Can we understand?

Cases of delusion often have the quality of a parable or fairy tale; of 'Once upon a time...' They are peculiar, cryptic, their meanings encoded. As with fairy tales, the themes inside these little stories are perennial: God, money, love, power, the reversal of fortune, death. Delusions are an imaginative space and people experiencing them appear to go through the looking glass into alternative universes, like Lewis Carroll's Alice climbing into Wonderland. When you pay closer attention to accounts of delusion from the past, however, you sense that there is something else at work. A delusion begins to seem more everyday and pragmatic – a psychological survival technique in action. Delusions may look like a retreat into the fantastical but in a key sense the opposite is true. These are not flights of fancy away from reality; they are a strategy to deal with reality. Unlike fairy tales, delusions are for grown-ups.

Introduction

A delusion is broadly defined as a fixed, false idea, not shared by others, unshakable in the face of decisive evidence contradicting it.

Where do delusions come from, and what do they mean? After all, it requires a considerable amount of imaginative work to create an alternative reality, and then heroic efforts to keep that reality going in the face of others – that is, everyone else – who *don't* share the same belief; who might even laugh at you, for having glass feet or a glass rear end. It's an intriguing question.

It's not a static situation; we've changed how we've thought about them over the centuries. In the classical world it was an imbalance of 'humours', later demonic possession, then organic brain disease.

A more consistent feature is how subjects cling onto their delusions, seemingly for dear life. What do delusions offer that is worth the trouble? What kind of help or protection?

Delusions are only just starting to emerge as a field of study in their own right. Accounts over the centuries offer us peepholes into this historically overlooked area of everyday human experience. They were typically written up as 'curios', or marvels of the mind, but there's more to know about these individuals in context, following each path through their daily lives; along the streets they walked, coming in and out of the shadows, glimpses of real human lives, struggles and powers of imagination. Can we spot traces of the route taken, how each managed to make a living, navigate love affairs and marriages, the birth and death of children, illness, wars or political or religious disorder, their sense of the future? Will their delusions become more understandable?

Looking back across centuries of experiences, we can see a common thread; common enterprises. Certain delusions seem to function in the same way. Organising ideas emerge. One persistent theme is lives that have gone catastrophically wrong. Delusions here are helping to assimilate a dramatic fall in status and to reconcile with the wretched existence which inevitably follows. The stakes for each of them in their delusions are as high as they come:

A housewife from 1920s Paris believes her husband and children have been replaced by substitute doubles. An Englishwoman in her Sunday best says she was swapped at birth and is the rightful heir to George III. She travels to St James's Palace to confront him with a petition and a butter knife. A man says he's Napoleon and barks his orders – countless lives depend on compliance. 'Madame X' calmly explains that she won't be needing supper because she is, regrettably, already dead. These people have never met, they are separated by hundreds of years, but they begin to talk to each other and reveal a certain solidarity. They have all experienced a reversal of fortune.

With other cases, delusions are a way of reconciling the irreconcilable. Francis Spira and James Tilly Matthews are men from very different eras. Spira is a lawyer living through the Counter-Reformation, Matthews a British diplomat and suspected double agent in revolutionary France during the Terror. Both are tormented by impossibly conflicted feelings and demands made on them. In Spira's delusion, God resolves things once and for all: Spira is damned for all eternity. Matthews clears up the confusion which surrounds him by identifying a political conspiracy: the British government is in league with the revolutionaries. There are good players and there are bad players and Matthews's paranoid delusion places him firmly on the side of justice.

And then we have our French glass king and his fragile posterior. The chaotic Charles VI responds so strongly to a new technology that he melts himself into it. Delusions of the body are often exquisite metaphors, witty and poetic, even when they result from grisly trauma, as with our clockmaker who survived the decapitation spree of the Terror but believes that he 'lost his head' under the blade of the newfangled tech of the guillotine.

And what of the people who want to know about delusions, who get obsessed with a cure? Who are they? The characters of the pioneering doctors we meet along the way are no less compelling, or confounding, than their subjects. Making his mark in the Paris medical journals in the early 1920s, for example, is Gaëtan Gatian de Clérambault. He gave his name to a delusional syndrome, otherwise known as 'erotomania', where a person believes that someone of high rank is in love with them (when they're not). While his theories circulated between the most distinguished Parisian doctors, he kept a personal fetish for silk and other sensual materials completely secret, from everyone except the mannequins in his apartment. He killed

himself with his First World War service pistol, staging the
desperate tableau of his suicide in front of a mirror as if it were
the final shot of a motion picture. The pioneers of new thinking around delusions in the nineteenth and twentieth centuries
were frequently traumatised by a direct experience of war. The
diagnosing doctor and the delusional patient undertake months,
if not years, of conversation. They engage in a dance of sorts,
and doctor and patient frequently have more in common than
they might imagine.

I spent many days eavesdropping on conversations from centuries past between physicians and the people describing their
experience of delusions, via the case studies which resulted. The
case notes are often brief sketches, with gaps and omissions,
and they are inevitably coloured by the psychiatric gloss, or the
religious or philosophical parameters of time they were written.
Many, of course, were recorded well before the language of
psychoanalysis was formulated. I've taken a certain licence in
passing a psychological lens over the oldest stories. My hope
is to try to understand a little more about the individuals who
hide behind the pseudonyms that headline the landmark cases
and how their delusions functioned in their lives; to flesh out
the elusive characters like 'Madame X' who believed she had
already died, and 'Madame M' with her stolen children and a
double for a husband. What's it like to experience an extreme
delusion first-hand? What were the specific experiences their
delusions answered?

We receive our subjects refracted through the mind of their
doctors and chroniclers with agendas, even delusions, of their
own. Occasionally, like tuning an old radio, you catch snatches

of what feels like an authentic voice in the static. Then they tell us how common our troubles are and have always been.

Don't mistake these stories for a collection of dusty curiosities from a long-lost past. Large-scale epidemiological studies in the 1980s and 1990s in the US interviewed members of the public, selected at random, to determine the prevalence of certain psychiatric conditions. This was the first time people who had not found their way to a clinical setting had been assessed. The findings surprised researchers. One such landmark study conducted in Baltimore in the US in 1991 noticed something interesting: 'The issue is the unexpectedly high prevalence of reported hallucinations, delusions and other bizarre behaviours among individuals who do not meet the criteria for diagnosis.'[1]

Historically, only the most extreme and bizarre cases were recorded, because these represented the people admitted to hospital. A far larger body of the people who were experiencing delusions remained under the radar, because the majority never required treatment and were otherwise relatively high-functioning. Simply put, we are all somewhere on the delusional scale: we all have at least one fixed, false idea about ourselves which other people, who know us well, would dispute. They could offer us plenty of evidence, too. Delusions are extremely common in the general population and, presumably, always have been. A modern reading suggests that delusions have always been closely tied to a person's sense of self, their views of the world and what is happening in it, and we should take them more seriously. The historian of madness Andrew Scull calls delusions a 'reminder of how tenuous our common sense reality seems to be' and this is an uncomfortable idea to sit with. Still we're drawn to them. Why do so many of us go to the trouble of creating these curious alternative realities? Which of our own beliefs might be false? Maybe the distance between the delusional – safe on the other side of the window – and the rest of us is no wider than a pane of crown glass.

Delusions offer rare access to private motivations, into the secret minds of others. We can't know on a day-to-day basis how other people are within themselves but when a person maps out their delusion for us we glimpse a whole world, designed entirely by them. We might be allowed inside for a tour, but only on the condition that we play by the rules as they set them out. It's invitation only and the by-laws of the land are non-negotiable. That is where they live. On the other side of the glass.

We have to feel our way carefully around to let our eyes get accustomed to the dimly lit rooms. As we listen to each story and become accustomed to the unfamiliar backstreets, not just of the *past*, but of an *alternative reality*, we begin to make out a living, breathing person and catch intimations from them of real-world and perfectly ordinary troubles and ambitions, imaginatively packaged for us to interpret. There are 'encoded hopes and possibilities' here somewhere, as psychoanalyst Adam Phillips has it. The difficulty is hearing the hopes and possibilities in what sounds crazy.[2]

Each of the individuals featured in this book laid a whole lot on the line by challenging reality so publicly; it follows they have something they want to say, commentaries and strategies to smuggle through. People with delusions have had their quieter, more sober messages drowned out by professional arguments over how they should be classified or cured. The longer you sit and listen to their accounts (they were amazingly good company during the lockdowns while writing this book) the more reasonable, even ingenious, they seem. Can we understand?

PARIS — ÉTABLISSEMENTS STOCKMAN — 150, RUE LEGENDRE

Postcard, 'La Couture'.

CHAPTER I

'Madame M' and
'The Illusion of Doubles'

3 June 1918. Paris. A forty-eight-year-old woman sweeps into a police station in the busy 15th arrondissement on the Left Bank of the Seine. She is dressed in the flamboyant style of the belle époque: all feathers and fur trims; performative hat and corsetry. She requests urgent assistance, and petitions for a divorce. The reason? Her husband, she insists, has been replaced by a series of doubles. Numerous 'lookalikes' have substituted themselves for other members of her family, her wider social circle, even for herself. There's one who's taken up residence in her apartment. In addition, she informs the desk officer, abducted children are being held captive in the cellar of her house and she can hear them calling out.

A couturière before she married, she will outline to her doctor, with an insider's eye for detail, what she likes to wear on a daily basis. In his 1923 landmark paper about her, her doctor lists: 'a black and lavender suit, a black "Amazonian" hat' (which seems to be the love child of a bowler and large fedora, a sort of female cowboy hat) 'with a veil and another hat in lavender'.[1] She doesn't say what kind the second is – perhaps she alternated between two different sizes for different occasions or seasons – but whatever the particular accessories that summer's day in 1918 her tailoring would probably have been theatrical, a signature of the era, creating an S-shape silhouette, making a shelf of bust and behind.

There is nothing unusual in a fashion-conscious member of the bourgeoisie flaunting her adherence to the most up-to-date mode, except when you consider that, by 1918, the belle époque is well and truly over. It's been over for years. The war that brought it to an abrupt end is still staggering to its own conclusion not far from where this peculiar scene is playing out. She is a walking anachronism. She continues detailing her day-to-day styling as evidence of her true identity, with forensic specificity, right down to the buttons and species of fur; features she will request that her doctor carefully notes down, so that no one mistakes her for her substitutes or misses any telltale alterations made by the impersonators. The real her, she points out for additional identification purposes, is usually accompanied by a little blonde girl 'in an embroidered linen dress, with her hair in a French plait, a white Brandenburg coat with ivory buttons and lined with duchesse satin, a straw cloche hat with a fancy white feather, yellow knee-length boots. In winter: dressed in a fluffy coat, a velvet cloche hat with white or beaver fur.'[2] The doctor will later share these details in his paper all about her. We can only imagine how this exotic creature with her stories of substitutions and kidnappings played to an overworked, nicotine-stained police commissariat in the middle of a city subdued by a seemingly endless war.

'Madame M', as she will be called when the doctor presents her case to the world, has intimate technological knowledge of the tailoring she describes. As a younger woman she earned her living piecing together fashionable garments for her clients. From the moment we meet her buttonholing a policeman it's clear she is a figure living in her own private reality. She begins to map out the sinister topography of this reality to an audience for the first time. Her colourful and frivolous appearance is a good way beyond nostalgic. She is a ghost of the relatively recent but already misty past. The belle époque commenced at the end of the Franco-Prussian War in 1871 and finished at the outbreak of the Great War. It was a moniker given retrospectively, meaning, essentially, 'the good old days'. The woman at the police station is wearing the motto proudly. She operates in her own youth: the Paris of haute couture, when clothes could afford to be costume.

Back at the police desk, the officer faces her. Our 'Madame M' has described the plot in headline terms, but she has much more she needs to get across to the authorities. So what does she want? Her allegations are ludicrous, *paranoid*. By the way she is speaking, agitated, desperate, this is not a game. The stakes could not be higher. Children are trapped and in danger. Identities and fortunes have been stolen. Is anyone listening? Will she get what she demands?

This feels like the opening scenario of an Edgar Allan Poe short story. Her allegations set up a dark and fog-shrouded mystery. Delusions, though, have always been a murky brew of real life and imaginative fantasy. Delusions are by their nature useful to fiction because they so naturally and economically dramatise the unconscious desires of a protagonist; they make the hairs on the back of your neck rise. You recognise something in what they are communicating about the trials and tactics of being alive, but you are not quite sure what. They are uncanny, just beyond your comprehension but deeply suggestive. So it is with the idea of the substitute double. The eponymous William Wilson in Edgar Allan Poe's short story of 1839 is a man plagued from his school days to adulthood by a rivalrous double. Poe portrays a man attempting to get away from the undesired aspect of himself by manifesting another self altogether, a splitting of the self. So often the fictional representations predate the clinical descriptions: Poe's story trumps Freud's concept of the repressed, unconscious alter ego and its drives by at least half a century.

In Dostoyevsky's 1846 novel *The Double*, the protagonist is a low-ranking office worker with poor social skills persecuted by a double who resembles him in almost every single detail except that the double possesses the very social skills that the protagonist lacks. He is tormented by this 'doppelgänger' to a denouement which sees more and more replicas present themselves, and is eventually carted off to an asylum.

The doppelgänger, literally 'double-walker', was portrayed historically as a ghostly phenomenon, and, more often than not, a portent of bad luck. The concepts of alter egos and double spirits have appeared in the folklore, myths, religious ideas and traditions of many cultures throughout history. In the mid-nineteenth century English speakers began to overlay this German word doppelgänger onto the concept of ghostly 'wraiths'. The English novelist Catherine Crowe's 1848 work *The Night Side of Nature* popularised the term. Crowe's book investigates a series of phenomena lying outside of scientific understanding, and she devotes an entire chapter to breathless accounts of double sightings from the eighteenth and nineteenth centuries. Most of her accounts are taken from cases shared with her by German physicians, each of which feature people who have been witness to doubles, either of themselves or of loved ones. Some can open doors and gates, some are seen by more than one person. They usually appear just before the death of the person who has been duplicated. In Norse mythology a 'vardoger' is a double who is witnessed performing a person's actions in advance. The doppelgänger is a version of the 'Ankou': a personification of death in Breton, Cornish and Norman folklore. Capgras's paper on 'Madame M' will be given the title of 'L'Illusion des "sosies"'. 'Sosies', or 'doubles', is a French word derived from Plautus's play *Amphitryon* in which the god Mercury assumes the appearance of Sosie, the servant of Amphitryon, connecting once again the clinical language of delusions all the way back to the myths of classical literature.

The psychoanalyst Otto Rank, a member of Freud's circle, introduced the idea of the double to psychoanalysis in his 1914 essay 'Der Doppelgänger'.[3] Freud's essay 'The Uncanny' followed in 1919 and suggested that fear of the doppelgänger was a symptom of an unconscious, repressed, fear of death. A double scares us because it forces us to recognise an anxiety we have repressed. It feels familiar.[4]

Rank made special reference to *The Student of Prague*, a German silent horror film of 1913. Set in 1820, the film is loosely based on Poe's 'William Wilson'. The image of a poor student is stolen from his mirror by a sorcerer who once promised to help him with his love match but instead becomes his double, and his love rival. The duel finale sees the student shoot this double, and, in so doing, kill himself. At the time of Rank's essay, the cinema was still a very new technology for a general audience, and its silver screens offered a kind of double dream world in themselves,

making them the appropriate place for audiences to watch a story
about the crisis of the individual play out. A few years later we see
the reflection of these screens in 'Léa-Anna B's eyes and her false
belief – the first formally described case of 'erotomania'– that
a complete stranger was in love with her. The overblown true-
love stories popular in 1920s cinema were a powerful two-way
experience for a person looking up at them from a dark theatre,
and encouraged dreaming and the projection of self onto others.

We don't know precisely what the desk officer at the commissariat for the Necker district made of 'Madame M' and her
startling claims. We do know, from the paper published about her
case that she was promptly escorted to the Infirmerie Spéciale,
a hulking edifice on the Île-de-France in central Paris housing a
network of public offices connected to the Paris police. Here she
underwent a psychological assessment, then was moved to the
Sainte-Anne mental hospital for more interviews, and then, on 7
April 1919, transferred to Maison Blanche, another pioneering
Parisian asylum of the age, one of a number established in the
aftermath of the Revolution where modern psychiatric practice
was first developed. It is apposite that our first case study takes
Paris as its setting, as the story of delusions and the part they
play in the history of modern psychiatry, is, in many ways, a
French one.

'Madame M' was at the mercy of the system now, but what
she found at Maison Blanche was not all bad. Here was someone who was not only willing, but keen, to listen to what she was
trying to say. The doctor recalls clocking her as she walked in,
describing 'the pretty costumes of her younger days' and 'a touch
of coquetry', apparently intrigued by this creature from another

age from the instant he first encountered her. The physician who set about taking down the details of this intriguing new case was a certain Joseph Capgras, and his name would be for ever associated with her disorder. Through her, he would become the eponymous father of a delusion 'type'. 'Madame M's symptoms defined it. Capgras is almost the same age as the woman in front of him, just three years older, both of them in their late forties at the time of this meeting, and the fame and reputation of both parties will be inextricably linked. They are from the same generation, but you wouldn't know it to look at them: Capgras wears his physician's whites over a sombre three-piece suit, arms crossed as he listens, a neat moustache beneath a penetrating scowl of concentration. He gives his subject a pseudonym, abbreviating her name to 'Madame M', which renders her a poster girl for the new category of delusion, disconnecting her from her real-life family history even as it immortalises her. Capgras follows a standard physicians' ethical code in not identifying a patient in the written case notes – he will only ever give us the first letter of her maiden name, 'M', and later drops in her married name, 'C'. He would have known both her given names in full, of course, but she is no less mysterious to him. Anyway, she insists that she wants nothing whatsoever to do with either her so-called maiden name, or her married name for that matter, because they are no more than evidence of abductions and substitutions. The question of her true identity, denied and stolen by others, is the root of her crisis.

Capgras's life trajectory has, like everyone's, been diverted by the war. In spring 1919 he has only just resumed work as chief psychiatrist at Maison Blanche, the institution to which

'Madame M' has transferred. He is fresh from a lengthy period of mobilisation, which began in 1914 when he was assigned to help the evacuating casualties at the Battle of the Marne, one of the bloodiest of the war. This first meeting with 'Madame M' is taking place in the theatrical surroundings of Maison Blanche, an imposing classical-style building in the shape of Louis XIII's chateau at Versailles. It is purpose-built as an asylum with three wings in a U shape around a courtyard. The asylum was evacuated during the war and turned over to a military hospital, and its hundreds of patients have also only just returned. We are just the other side of the conflict, but the impression of the war is everywhere.

Capgras will observe 'Madame M' for several years searching for clues as to the cause of her extraordinary beliefs. After countless audiences, collating any biographical scraps and domestic details which might make sense of her, he commits his thoughts to paper. He will share his appraisal of 'Madame M' in a lecture at a meeting of the Société Clinique de Médecine Mentale, published as 'L'Illusion des "sosies"', or 'The Illusion of "Doubles"'. His peers are all interested in the newly revived subject area of delusions and love nothing more than a chance to argue out their different theories and interpretations in public. Capgras reveals the case of 'Madame M' to his audience with a dramatic flourish. 'We present here', he says, 'a paranoid megalomaniac... interesting due to the existence of a delusion, or rather a strange interpretation: for about ten years she has been transforming everyone in her entourage, even those closest to her, such as her husband and daughter, into various and numerous doubles.'[5] The principal delusion, the doctor says, is substitution. He also lays out a subplot for their interest: the 'illegal confinement of a large number of people, particularly children, in the basement of her house, and throughout Paris...' The theatricality doesn't stop there. Unbeknown to the audience Capgras has brought 'Madame M' in person to the lecture theatre and she is waiting offstage.

Capgras declares this delusion an 'exceptional case...rich... fantastic and yet systematized'. He has clearly enjoyed the challenge of deciphering 'Madame M's messages through the clues. Why has she committed herself to such an absurd and palpably false theory, a theory which flies in the face of the evidence? What has happened to her? What does she want? We might also ask why is Capgras so interested in this one particular woman that he devotes so much attention to her and her wild accusations? The audience wait expectantly for him to unroll his theories and interpretations. What facts has he managed to establish? The waypoints of her life, the ones he's been able to elicit and thinks are pertinent, are these:

Mme M is now 53 years old with no family history of psychiatric disorder. She had typhoid at the age of 12, received a primary education until the age of 14, then learned and practised the trade of couturière. Married in 1898 at the age of 29, one year later she had a son who died (she believes he was substituted), then she had twin girls, one of whom died (abducted according to her), the other is in good health and now 20 years of age, and in 1906 twin boys were born, both of whom died at a young age (according to her one was abducted, the other poisoned). She lived comfortably, her husband owning a large dairy business. A restrained, sober person, the only thing she

ever took to excess was coffee… In the ward Mme M is usually
calm, polite, even kind when one is not talking to her about her
delusion; she does not make any friends, remains completely
idle, and refuses all work. She gets annoyed at the name M;
sometimes she writes long letters, sometimes also she indulges
in soliloquies accompanied by gesticulations which bear witness to her intellectual excitement. This is noticeable when
the patient exhibits her delusion, which she does in a rather
complex manner, talkatively, verbosely and with an extreme
flow of ideas which require precise, close cross examination to
stem her natural tendency of continually diverging from the
point… Mme M, has never been aggressive, but has made two
attempts to escape, and has escaped once.

After a no-nonsense opening, signalling a fairly ordinary background, the allegations are stark and disturbing. Infant abduction
and poisoning. The picture is contradictory, even when painted
in the brusque manner you would expect from a doctor's working
notes. She is not a couturière any more, she is a married woman,
living comfortably as the wife of a successful business owner.
Four of her children died in the early years of her marriage,
her doctor says, although she flatly denies this. She changes
quickly from calm and polite to displays of irritability, jealousy,
grandiosity. She's prone to physical agitation as she tries desperately to communicate her story, but she is also lazy, preferring a
totally solitary existence. Cumulatively the characteristics begin
to move, like pieces inside a kaleidoscope, and make an image
of 'Madame M' difficult to see in the round. Capgras notes her
respectable marital status and comfortable financial position

along with the listlessness. His aside that her only vice was a taste for coffee sounds quaint now, but then comes a reference to her attempts at escaping the hospital. We remember that 'Madame M' is in an asylum. She has lost her autonomy. She's a prisoner. She's not the first of our subjects to try to escape. A hundred years before, Margaret Nicholson, the housemaid adamant she was George V's rightful heir jumped over the wall of Bedlam in London. Nicholson was caught running to her brother's pub for sanctuary. Like 'Madame M', she was returned to her cell.

Capgras continues presenting his case to the assembled crowd of peers, whose rapt attention he commands. He cues the big reveal. 'Madame M' herself is brought into the room. She is now wearing the plain asylum uniform and she is told to stand in front of the phalanx of doctors and retell her story, live, for their consideration. Psychiatrist Gaëtan Gatian de Clérambault is among their number, watching with interest. He is chief physician at the Prefecture of Police, with its psychiatric emergency room, the Infirmerie Spéciale where 'Madame M' was brought for assessment. He may well have been there when she was first brought in by the police officer, and could even have conducted the initial evaluation. She would certainly have turned heads at the Infirmerie. He'd remember her. Paris's underclass wash up there day and night – it is a place where criminality, destitution and mental disturbance congregate: absinthe drinkers, prostitutes, homeless, poor, the delusional. It provides a clearing house for extraordinary stories of all sorts. De Clérambault is a former classmate of Capgras, just a year younger, and a rival in the psychiatric avant-garde. Only a year or so earlier, he was the first formally to describe (and attach his own name to) another 'type' of delusion: 'De Clérambault syndrome', or 'erotomania', holding up 'Léa-Anna B' and her belief in George V's undying love as the example.

Born in Verdun-sur-Garonne in south-west France, Joseph Capgras completes his medical training in the hospitals in nearby Toulouse. He gravitates to psychiatry as a speciality under the influence of his uncle, an asylum doctor in the Paris area, and comes top in a series of highly competitive examinations which leads him to a post at Maison Blanche. He is an ambitious man, who has made his way from the provinces to Paris and at this point he is already well known for a 1909 work on psychosis and disorders of recognition, *Les Folies raisonnantes*, co-written with his mentor, Paul Sérieux. Sérieux also collaborated with German psychiatrist Emil Kraepelin, a giant of psychiatry, who first described 'dementia praecox' – the condition which became known as schizophrenia – as a disease of the brain. It was Kraepelin's way of seeing things that cast delusions out to the margins. They became minor symptoms of his new disease and had remained in the wilderness for many decades, considered unworthy of research as discrete subjects.[6]

The lecture continues and de Clérambault jumps in to offer his tuppence worth. He suggests 'Madame M' might be hallucinating. 'Madame M' has been on stage listening attentively. She interrupts in turn, 'entering vivaciously' into the discussion. She's sticking to her story and she repeats some points for those at the back. She is 'not a woman', she is a 'young girl' and lives depend upon someone taking action now.

Interwoven with her delusion of substitute doubles are what modern psychiatry would call 'persecutory' and 'grandiose'

delusions – two more of the principal categories of delusion clinicians will establish later in the century. She points the finger at
the many people who are out to get her, identifying coordinated
plots between officials, doctors and administrators. Three or four
years after their marriage, after the death of their twin boys in
1906, her husband noticed a 'nervous' state develop in his wife,
followed by 'jealousy' and 'delusions of grandeur'. Another strand
in 'Madame M's story is the question of her own noble lineage.
It is quite a family tree:

> 'I am from a very important family,' writes Madame M; 'I am
> the granddaughter of Princess Eugenie; I was born to the
> Legion of Honour; my father was the Duke of Broglie and
> my mother Mlle de Rio-Branco, the daughter of the Duke
> of Luynes.'

Capgras teases out key facts from the genealogical tangle for his
paper, sometimes quoting her, at other times speaking for her,
summarising to move things along:

> Her paternal grandmother is the queen of the Indies. De
> Rio-Branco is the name of the children of Henry IV from
> whom she descends; she is a relative of the Duke of Salandra.
> She adds, 'M. Pierre-Paul M, who died at my house, certified
> that I was not his daughter, that he had acted criminally in
> hiding me from my parents and that I was 15 months old
> when the abduction was committed.' Therefore she believed
> she had been substituted for the daughter of M. M. from
> the cradle.

Now Capgras glosses the money situation according to his patient:

> She claimed her fortune to be immense, that all of Rio de
> Janeiro belongs to her grandmother who owns considerable
> mines in Buenos Aires. 'I am certain,' she says, that I have
> been left 200 million francs by my grandfather, Louis XVIII,
> who lived at the Tuileries... Since my childhood I have been
> pursued by a gang who knew about my wealth, since I was
> taken away from my parents and abandoned with a M.M.'
> Thus substituted for the daughter of this man, she should
> not be called 'M' but Louise C., the name of her husband, or
> Mathilde de Rio-Branco, the name of her true family.

She is insistent. She was abducted and swapped as a fifteen-
month-old baby and is now pursued by the thieves who know
the truth. She renounces any ties to the family who took her in.
'Pierre-Paul M.' is the name of the man people call her father.
He died at her house, she says, in a throwaway comment. These
seedy and oddly specific domestic details jar with all the grandiose
talk about Louis XVIII and Versailles and the fortune. It is while
she is in full flow that her own first name slips out. She is called
Louise. She's not remotely interested in owning the name, of
course, this is not the identity she is asserting, but now we know
it, and it is a touching detail: informal and ordinary against the
affectation and melodrama.

The melodrama starts up again: 'My character has never been
anything other than upright... I am an untarnished woman...
My signature has value.' 'Madame M' wants to stress that the
real her, Mme de Rio-Branco, is 'honest' and 'of sound mind'.

'Just think of the trap in which I find myself; one must be gifted with an intelligence and an uprightness that will match up to each and every test, to hold one's head up high to all the villainous leeches. So there is a story of a true French Woman who wanted to save three-quarters of the Universe,' she says, not afraid of hyperbole. 'I would have done great things with this fortune. I would have done some good for everyone.' In her stagey patriotism we hear echoes of the 'Napoleons' who belted out their orders along the neighbouring corridors many decades back, and even the 'Christs' who turn up in Israel. Delusions of grandeur bestow terrible responsibilities and sacrifices as well as riches. She would save everyone with her bountiful generosity, if she could just have her true identity back. She expresses frustration at her own powerlessness, but even this is underpinned by her grandiosity. She is powerless because the whole world is rigged against her. It's a nightmare and she's at the centre, with everyone giving her downfall their full attention. The gang of 'forgers' and 'swindlers' who stole the money of her birthright have even tried to poison her food at the asylum by putting arsenic in her 'spices, her food and drink'.

Her case brings together some of the most enduring themes of delusion. Capgras is careful, though, to foreground her main theme again: the disappearance of people close to her and their substitutions. The doubles have stolen her marriage certificate as part of an elaborate identity theft, so that she will be punished for the crimes of these substitutes. This identity theft has entailed surgery which wouldn't be out of place in a gothic Hammer horror film:

To clarify her identity, and complete her justification, she points out the transformations of which she has been the object. 'I was blonde, they made me chestnut, with eyes three times the size; they were rounded in front, now they are flat: they put drops in my meals to take away the features of my eyes, and the same with my hair; as for my chest, I no longer have one…and that's why no one recognises me anymore…

Capgras is a man interested in motive: after the war he was appointed a forensic expert to French courts of law, and he examined numerous criminals, publishing 'Crimes et Délires Passionnels' in 1927.[7] The question of legal responsibility was a particularly thorny one at the time and Capgras argued that the concept be replaced with notions of psychological 'morbidity' and 'noxiousness'. He liked to solve complex cases, to decode psyches. In 'Madame M' he had much to unpick. And then her husband joins the story, stepping centre stage. The melodrama becomes more macabre:

Her husband M. C. also disappeared: a double took his place; she wanted to get divorced from this double; she drew up a petition and made a request for a separation to the courts. Her husband had been murdered and the 'men' who came to see her are doubles of her husband; she counted at least eighty.

She ups the ante again. Her husband wasn't just swapped out and disappeared, he was murdered. He was killed before he was substituted. A stream of impersonators visit her. The circumstances are shady but her grounds for divorce are clear. She must be allowed to separate:

> Moreover, she continues, if this person is my husband, he is more than unrecognisable, he is a completely transformed person. I can assure you that the imposter [sic] husband that they are trying to insinuate is my own husband, who has not existed for ten years, is not the person who is keeping me here.

She alters her allegations slightly: there is one individual taking the place of her husband but he is nothing *like* her husband. The cover-up is obvious and unconvincing but the authorities deny all the evidence. This so-called 'husband' is a stooge, claiming rights he doesn't have, keeping her prisoner.

Capgras tells us about another particularly unnerving claim. She says that because of the abduction of her son, and his substitution with a double she was 'thus at the burial of a child who was not mine'. The substitute 'was poisoned at twenty-two months. She noticed this by looking at his nails.' This specific detail of the appearance of the deceased child's nails has the ring of grim experience, even though it's set within the far-fetched scenario of a funeral for a child who's been switched.

She returns again and again to her physical appearance and her clothes. Capgras reports her confused, and confusing, explanation for her altered features:

> So that there is no longer any mistake about it, here are my personal details, for which there is urgent need because of the change which has taken place in me during the last

twenty-five years [...] I have been transformed in order to completely change my person. Being unrecognisable, apart from a few identifying marks, it would be possible for them to pass me off as an insane person. My particulars: I was blonde, brown-eyed with some black areas in the brown, scars near my right eye and various others, a scar on my right hand and a turquoise ring, which was confiscated from me, and I had two small freckles on my neck. [...] This person, who is me and whose true particulars I am relating, is honest beyond doubt... No mistake can possibly be made, I am the only person with these distinguishing features. She cites also her surname, all her Christian names, her full date of birth, and the full address of the person that she replaced following her substitution. Her Paris address includes the arrondissement, the quartier, the street, the number, floor, and even the side on which the flat is situated. In short, she is concerned not to be taken as a double for herself and because she notices the signs of age, or rather the signs of persecution, which have changed her face.

This woman is a stickler for accuracy down to the finest details and she is keen to help organise the clues for her doctor and provide all the administrative help she can. She's most keen to demonstrate the workings of the plot to steal her identity and replace her with a double. She overwrites the most ordinary personal details with evidence of this plot, even pointing out its influence and logic in the normal signs of ageing. Her persecutors have inflicted deliberate changes on her appearance in order to separate her from her true identity. They present her as mad so that she won't be believed and have worked to undermine her public credibility. She also picks out the various scars on her face and hand as identifying features, but gives no explanation for them. Her past is shadowy, and marks like this whisper of something sinister.

Capgras ploughs on with his lecture. How will he attempt to frame her delusion? Will we agree?

'Madame M' is scrupulously careful to namecheck the cobbled boulevards and alleyways of the 15th and 14th arrondissements. She is up and down them on her daily errands. The 14th will become even more famous by the time Capgras's landmark paper 'L'Illusion des "sosies"', or 'The Illusion of "Doubles"', is published. Capgras is fifty-three years old when it begins circulating and the *années folles* – France's 'Roaring Twenties' – are just kicking off. The cafés and entertainment spots of Montmartre draw influential artistic figures to the area which 'Madame M' knows so well, some American expats – Hemingway, F. Scott Fitzgerald – as well as Picasso and Cocteau. By the time the wider world gets to know her name, 'Madame M' is no longer at liberty to walk these streets and enjoy the atmosphere.

All we know for sure about 'Madame M's feelings for the family she grew up in is that she wants nothing to do with her family name 'M'. She even claims that her 'fake' father 'Pierre-Paul M' confessed to the crime of abducting her. She doesn't disclose anything more. The belief that you've been swapped, and are actually, for example, a prince or princess, rather than a member of an ordinary family, is not an uncommon fantasy in children. The difference is that most grow out of it. With her elaborate plot 'Madame M' gives us a strange analogue to reality, but a few details stand out within it because they are so particular, hinting at real-life family secrets, estrangements or unresolved conflicts.

'Madame M's childhood is brushed off with the mention of a bout of typhoid, but we can work out that she was born in 1870, the year the Siege of Paris began. Her birth coincided with the defeat of the French by the Prussians exactly forty-eight years before she met Capgras. Her social circle in her early years would have included previous generations who had felt the aftershocks of the Siege and the Paris Commune. They surely told the children the story of how a working-class revolutionary movement briefly ruled the city, refusing to accept the authority of the French government after the surrender to the Prussians. They might have been more cautious talking about the widespread starvation that followed. But it was the new prosperity of the Third Republic that 'Madame M' came to Paris for, and by the time she was there making her living, in the years before the *next* war, France had expanded its empire and was reaping the economic rewards. Earnings were good for the burgeoning middle classes. Before her marriage, 'Madame M' would have witnessed the confidence of the era at close quarters. Even if she couldn't afford the material trappings herself at the time, she would have seen the showiness of her age reflected in her clients, who flitted in and out of her place of work, looking for a dressmaker to realise a certain fantasy in cloth so they might signal publicly they were part of the glamour. The architectural landmarks of her daily stamping ground feature in her conspiracy and they are the shops and apartment blocks skirting Montparnasse, a stone's throw from the lively cultural centre of the belle époque.

There was little by way of welfare assistance for the poor and France had a large economic underclass which never enjoyed the benefits of the belle époque at all. 'Madame M' began work in 1888, and spent ten years of her life making what were essentially costumes, behind the scenes, before she married 'M. C.' in 1888. Nevertheless, begging, destitution, prostitution and illnesses from poor sanitation would have been impossible to avoid on the streets of Paris in the years leading up to the Great War. As well as typhoid, tuberculosis was the great threat to public health by this time. The mannequins 'Madame M' dressed up in the fashions of Paris and Vienna were styled to appeal to the new money in the city, and would have given no hint of the parallel experience. Her formative years as a young adult were spent in a world of make-believe: the Exposition Universelle of 1889 was held the year after 'Madame M's marriage. With its showstopping Eiffel Tower spectacularly illuminated as an entrance, the world's fair showed Paris off as the cultural centre of Europe, at the leading edge of technological and scientific innovation. Light entertainment sparkled in the cabarets; the saucy cancan whipping up enthusiastic crowds with high kicks. The rich could travel easily for the first time on the new network of railways to exotic resorts, and ladies' fashion magazines included advertisements for glamorous destinations like Biarritz or the Italian Riviera, intended to catch the eye of the bourgeoise wife with suggestions of wanderlust. 'Madame M' could not have avoided material decadence, conspicuous consumption and aspiration any more than she could ignore the poverty and sickness. This was a divided society and anyone attempting to live and succeed within it had somehow to accommodate these contradictions.

One landmark plays a starring role in her alleged conspiracy. It lay below the street level of her neighbourhood: the Paris 'catacombs'. The 14th arrondissement boasted an elaborate network of underground tunnels, two hundred miles of passageways filled with human remains. This was a counterpart world, only feet away, but inaccessible and mysterious. Over a century earlier, just before the Revolution, the city authorities had co-opted a number of abandoned quarries as a vast ossuary. Labourers were to transfer skeletons there from the overflowing cemeteries at the edge of Paris. 'Catacombs' was the name given to a series of underground tombs along the Appian Way in ancient Rome. Digging the tunnels in Paris centuries later created an artificial hill of rubble referred to by students fond of a good classical allusion as 'Mount Parnassus' after the mountain in central Greece celebrated by the Greek poets. They recited poems at its foot, and so Montparnasse got its name. 'Madame M's crisis played out alongside the First World War and its aftermath and the creation of the catacombs was well outside living memory. Even so, the network was a reminder of another time when the city was overwhelmed by its dead. It is, then, a poetic setting for her delusion, whose principal image system is that of doubles, of mirroring, of a daily life shadowed by enemy networks and impenetrable motives. Beneath rue Mathurin-Regnier, a street she knows, is another world, packed with the desperate, the abducted, and more substitutes:

> ...there are dungeons, an artesian well and vaults where about 28,000 people have been shut up since 1911, a group of individuals strip the people of everything they have and shut

them up in the cellars: they were in correspondence with a tenant in her house. The 'double' of a Madame P. Beneath the Military School, l'Avenue Suffren, le boulevard Dupleix, rue Dutot, about twenty children are asking her to get them out. Under her own house she can hear the voices of children calling out: 'Mother, I beg you, come and get us out.' Beneath it's a complete system of 'amphitheatres', of 'underground passages'; the people have gone underground as if by means of a service or goods lift, by stages, by stages, and someone is doing away with them. Living people are being buried in catacombs. Near Pasteur thousands of individuals have been shut up and mummified. The cellars of Paris are full of children. During the war, because of the planes which were flying above the house, many people including children went down into the cellars and did not come back up; they found that they were trapped. Underground operating theatres were set up to disfigure people and it was said that they were people returning from the war. The shelters are not very safe, because of those who go down, very few come back up; she forbade people from going down. 'The German fighter planes are firing blanks; there are no bombs, people are wrong to seek refuge in the cellars: many young girls are unable to get back out, as the opening is blocked up. The Métro is fatal for us, because the French and English armies have been put down there: the crisis of strength in the military arises particularly because of the disappearance of regiments underground, in the Métro... More people have disappeared like that' – she writes – 'than have been taken prisoner.' The military uniforms loaded onto the trucks belong to soldiers undressed underground.

Her doctor gives us 'Madame M's sketch of what's going on beneath her feet. Here are circles of an underground hell worthy of Dante's *Inferno*. This subterranean nightmare is teeming with the horrors of the First World War. The underground spaces of the 14th arrondissement are a prison for disappeared children and soldiers. The soldiers shout reports of torture up to her: 'Madam we have been underground for three years and they have been dragging us around like carts and chasing us with whips.' She believes that the war itself has been drafted as a cover for an ever wider-reaching covert operation. There are no bombs. If houses are demolished, it's just to confuse people. Germans are firing blanks. It's part of a scheme to get people into the underground prisons where they can be operated on, and from which few emerge. Doctors and nurses at Sainte-Anne and Maison Blanche are all in on it. They all have doubles, and they also disappear 'underground by means of a service or goods lift' to a 'system of amphitheatres'. Even her husband, the substitute version, descends under the hospital at one point on a visit to see her.

The image of military uniforms being loaded onto trucks stands out. Another very particular and mundane detail that has the quality of something she'd really witnessed; routine activity by the side of the road that stopped her in her tracks and haunted her afterwards. She mentions these uniforms only to explain them away quickly as more evidence of the plot: the clothes belong to the naked soldiers who are trapped underground.

She reminds us that the catacombs were no longer the only tunnels under the city. The digging out of the Paris *Métro* had begun in 1898, with the first line inaugurated in 1900 during the Exposition Universelle. On 4 November 1920 a new line opened linking Montparnasse to Montmartre. This was an ongoing feat of civil engineering and technology that 'Madame M' will have passed on the streets of Paris over the years. It stopped and started just as the catacomb excavations had. The catacombs, at twenty metres beneath the pavement, were far deeper than the sewer or *Métro* system, but the different strata of tunnels become part of 'Madame M's parallel universe, and the conduits connect as part of her far-reaching conspiracy.

The image of the soldiers lost underground will have been a potent one to Joseph Capgras, after his experiences as a field officer at the Battle of Marne in September 1914, a key Allied victory and one of the turning points of the war on the Western Front. Skirmishing had reached the outskirts of Paris. The city was saved by the victory, and France's position in the war preserved, but this had come at great human cost: with an estimated 500,000 wounded or killed in a single week, including 250,000 French, of which 80,000 died, and around 250,000 Germans, this was the highest number of daily losses of any battle on the Western Front. After their defeat, the Germans retreated, leaving more than 11,000 prisoners for the French, and it was some of their pitiful and traumatised number that Joseph Capgras assessed and treated at evacuation hospital No. 38, Section 5, not long before he met 'Madame M'. In 1916 Capgras was assigned to the psychiatric centre at Orléans, and, with Paul Juquelier and Joseph Bonhomme, provided reports on 'mental confusion', which had been precipitated by the war, a condition which would be interpreted by others as 'shell shock'. In July 1917 Capgras presented with his colleagues to the Société Clinique de Médecine Mentale with accounts of wounded soldiers' 'stupor', 'dreaminess', 'auditory and visual hallucinations'; 'cinematic' replays of battle, and the incurable catatonia of some, such as those who had experienced the battle at Verdun.[8] In August 1917, Capgras was promoted to the rank of '*médecin-major de 2e classe de l'A.T*' and in 1919 he was decommissioned and resumed practice at Maison Blanche. In 1937 he was made *Chevalier de la Légion d'honneur* in Dijon for his work.

It is not hard, then, to see why Capgras might be receptive to a woman whose tales of substitution and abduction offered further opportunity to explore links between trauma and mental disturbance.

Delusions often reflect the social preoccupations and anxieties of the age, and, given her delusion of doubles, it's striking that 'Madame M's lifetime to date had seen the technical development of the media of optical duplication and projection. The first cinematic public entertainments like the 'Phantasmagoria' (hugely popular in France from the end of the eighteenth century right up until true moving pictures arrived at the turn of the nineteenth) used shadows, waxed mirrors and smoke to create spectral illusions which were macabre in nature, intended to inspire terror and dread. These quietly unsettled many people's notions of individual integrity and uniqueness.[9] Capgras uses the word 'phantasmagoria' to describe 'Madame M's 'wild imaginings'. The technical possibilities of duplication in devices like the 'magic lantern' converged with wider spiritual and psychological anxieties. They dramatised the central question haunting 'the double': how can a person distinguish between what is truly an external presence and what is, in fact, an internal, psychological conflict projected outside of oneself? 'Madame M' had been brought up in a world captivated by the morbid, 'out of body' visions created by the magic lantern, which seemed to raise the dead from the shadows. By 1918, she would perhaps have visited a parlour or two to marvel at early true moving Kinetoscope cinema. It's hard to miss the visual influence of the Phantasmagoria in the nightmarish, hallucinatory visions of the subterranean prison she describes. This was a world dealing with the losses of the war. A concurrent fashion for seances channelled a collective yearning for the dead and the hope that they were still alive and might return in some form or other.

France saw rapid urbanisation after 'Madame M' arrived in Paris, and she navigated a city that was undergoing a population explosion. The war brought death on an industrial scale to a generation of young men, made possible in part by technological developments such as the machine gun. There was a stark new cheapness to human life and labour, but the technological progress of industrialisation had also created the city's prosperity. A person might lose their footing, sensing their place in the world, their power and value threatened, at the same time as noticing opportunities out there for the taking. They might choose to rebel against the demotion.

Capgras is fixed on the question of what laid the groundwork for 'Madame M's challenge to reality as most people generally accepted it. A hundred years on we are particularly curious about the woman behind the pseudonym, called Louise, a woman of whom we've had only fleeting glimpses. What was her everyday life like before she was institutionalised in Maison Blanche?

Capgras notes that 'Madame M' is most articulate when she's writing. Apparently she tends to ramble off the point when extemporising, but her doctor is impressed by the coherence of her letters and he quotes extracts liberally. He tells us that she received 'some education' until the age of fourteen.

Girls' education was limited in France, but primary education was compulsory from 1882 for both sexes in the country from the ages of six until thirteen, and included mandatory tuition in needlework for girls. Free, secular schools came in with the Jules Ferry Laws, named after the staunchly republican lawyer and Minister of Public Instruction in the 1880s, who wrested control of education away from the Catholic Church. 'Madame M' may have attended a Catholic school before that, paid for by her parents, but, regardless, she received good enough teaching, and was a sufficiently able student during her short school career to be the confident writer Capgras observes. The biographical details in the records include only a brief mention of her métier, stating that after she finished her education at fourteen, she went straight to work as a couturière.

Atélier Maison de Couture Storch, 28 rue Henrion
de Pansey, Paris, 14th arrondissement (postcard).

The job title is ambiguous. Before her marriage and the war, 'Madame M' may have been a sole trader, like the number who advertised in the back of the fashion magazine *La Couturière Parisienne*, offering her services as a dressmaker or making corsets for a competitive price. These advertisements sat among others for sewing machines or furs – '*Beaver or Canadian skunk, Madame?*' (in Paris, exotic feathers and furs were featured in fashion as never before). Still others promised beauty enhancement or new feats of vertiginous millinery – and even fortune tellers who also seem to have been in high demand, suggesting that anxieties about the future simmered on in the minds of a population who were enjoying only a brief window of play before the next major conflict. In an edition of *La Couturière Parisienne* from before the war, next to dress and embroidery promotions, a 'Madame Renault' offers women lessons in hypnotism and 'magnetism', a treatment that was continuing the legacy of the German doctor Franz Mesmer's theories of innate invisible forces as physical cure, passed down all the way from the eighteenth century. Mesmer's theories were still beckoning come hither to the general public: they had captivated the British diplomat James Tilly Matthews when he was in Paris more than a century before. Matthews gave these forces form, and malevolent function, in his magnetic mind-influencing machine the 'Air Loom' which was at the centre of his delusion that traitorous revolutionary conspirators were trying to overthrow the government.

A hundred years later, a couturière advertising in a ladies' magazine next to the likes of 'Madame Renault' and her lessons in magnetism generally offered tailor-made services, with a mannequin for modelling the clothes on, cutting from a pattern,

sewing and assembling, adjusting hems and adding embellish-
ments and accessories according to trends: feathers, embroidery,
buttons and gloves. Work like this would have been increasingly
time-pressured, as the couture industry moved to fast-paced
seasonal cycles.

Large numbers of very young women migrated to Paris from
the provinces to make a living, employing the manual skills they
had acquired at school, and the nineteenth century saw a five-
fold increase in the population of Paris, which reached well over
three million by 1896.[10] Having learned her trade, a dressmaker
might join a large atelier or couture house, even working her
way up to occupy a managerial role within one of these fiercely
hierarchical organisations, heading a section, or taking up a
particular speciality.

She might, of course, join a very low-paid collective of two
thousand or so seamstresses, or *petites mains* (literally 'little hands'),
backroom girls who painstakingly brought haute couture creations
to life in larger ateliers. 'Madame M' mentions burying a child
in Bagneux, to the south of Paris, so perhaps had links there.
Capgras mentions in passing that she went 'to the Auvergne to
find her children' so it's also possible she had family roots in that
particular part of rural south-central France, whose primary
industries were farming, cheesemaking and glassworking. Any
migrant to the city from a rural, provincial life would be faced
with the brazen pretension of Paris. It is not hard to imagine
mental conflicts emerging, 'cognitive dissonance', between where
a woman like her felt she should have come from, and where
she actually came from; what she could pretend to be, and what
she was.

The American social psychologist Leon Festinger presented the first fully formed ideas about how internal conflicts might motivate people in his *Theory of Cognitive Dissonance* in 1957. It's a work very much of the twentieth century but it's useful as a way of understanding delusions from the more distant past. Festinger's theory came from his observations while he was imbedded in a cult and living at close quarters with its members who were expecting the imminent end of the world. They had left jobs and homes, but when the prophecy failed to come true, rather than desert the cult in despair they set about recruiting additional members with renewed energy. Festinger concluded that they needed support for their beliefs more than ever, in order to lessen the pain of the blatant failure. According to Festinger's theory, a person who is holding onto contradictory beliefs, ideas or values experiences significant psychological stress when forced to challenge one of them by the outside world. Internal inconsistency is very uncomfortable, and people will do almost anything in their power to make things consistent in order to function well. To Festinger, this driver is even more powerful than emotion, habit or financial reward. The difficulty of sitting with contradictory positions, when each seems true, is resolved most directly by blindly believing *something*.

'Madame M' took a step up with her marriage in 1888 in terms of the greater financial security that came with it. A 'dairy business' probably means a retail operation rather than a farm, and husband-and-wife business enterprises were also very common. This 'murdered' husband whom Capgras abbreviates to 'M. C.' is a shadowy figure but we may imagine him crossing paths with his wife-to-be at the sales end of his dairy business.

The boulevards of the 14th or 15th arrondissements were strung
with boutiques, and stores offering daily provisions, like *laiteries*, as
well as trades servicing a booming economy, places where atelier
workers would be able to associate with wealthier merchants and
clients. Did 'M. C.' meet his future wife while sourcing clothes for
himself or a female family member? She was far more protected
as a married woman with a solid business funding the household,
a rung or two up the ladder, but did she miss the sovereignty of
life as a dressmaker, or the energy and comradeship of her work
in a collective of *petites mains*?

And then, in the early years of her marriage, she lost four of
her children. The death of children in infancy was, sadly, all too
common at the end of the nineteenth century, but for 'Madame
M' these years included a series of personal catastrophes.

Then comes the war, and the devastation converges with her
private loss. Capgras reports the husband's testimony – she was
mentally unbalanced for many years before this war started,
he says. It was over the four years of the war, though, that the
delusion really came to fruition. There was a crisis point in 1914
after which she began to complain about thousands of substituted versions of her own daughter plaguing her, as well as other
dreadful apparitions of children with 'stitch marks' on their faces
from operations to remove their thoughts.

One day obsessed by this notion of the abduction of children,
following the death of four of her own, Mme de Rio-Branco
no longer recognises her daughter; someone has abducted
this child and replaced her with another child who bears a
resemblance to her; the next day another little girl similar to

the one before appears; two days later there is another substitution of a double. From 1914 to 1918 she writes, more than
two thousand doubles of her daughter have passed before
her eyes; every day, indeed a number of times a day, a little
girl appears.

Et qui n'est chaque fois ni tout a fait la même, ni tout a fait une autre.

This fragment of poetry translates as: 'And each time she is neither entirely the same, nor entirely different.' It is not entirely
clear whether Capgras is quoting his patient here or providing
his own gloss on her delusion, but 'Madame M' has apparently
illustrated one of her many expository letters to Capgras with an
extract from a Paul Verlaine poem of 1866, 'Mon rêve familier'.
The poem would have been an old one, even then, but it communicates perfectly 'Madame M's strange sense of finding her
'daughters' both familiar and alien. The fact that you can't tell
if it's Capgras or 'Madame M' talking is an expression of how
close the two lives have become. The quotation is from a section
of Verlaine's poems grouped under the subtitle 'Mélancholia'
from his collection *Poèmes saturniens*. Verlaine was a member of
the Parnassian movement. Like Montparnasse, this group of
Parisian poets took its name from the home of the Muses, Mount
Parnassus. The Parnassians published only a few volumes, bringing back classical subjects and rejecting fashionable sentimentality
in favour of formal discipline and a cool, emotional detachment.
If 'Madame M' read Paul Verlaine, as it seems she did, then she
knew about melancholy, and she knew this poem well enough
to quote it by heart. She was sixteen when it was first published,
and already working in Paris.

Le « SECOURS DE GUERRE » au Séminaire Saint-Sulpice
L'Atelier de Couture.

Le Secours De Guerre au Seminaire Saint
Sulpice, 'L'Atélier de Couture' (postcard).

What was real life like for 'Madame M' during the earlier
stages of the war? She would probably have been drafted into
the war effort, just as other seamstresses and couturières were.
Women took new positions in the workforce, in munitions factories
and postal services or public transport. Anyone in Paris would
have been aware that the front lines were moving ever closer to
the city and would have endured shelling by the Germans, as
well as rationing, food shortages and the newest silent enemy, an
epidemic of Spanish flu spreading with terrifying speed across
Paris in the spring of 1918. This invisible scourge took hold as
'Madame M's crisis peaked.

Her delusion came fully to life during the last year of the
war. It is plausible that she did witness soldiers' uniforms being
loaded or unloaded in Paris at this time. Since the boots of dead

soldiers were never reused, it is most likely that the lorry she saw
was being loaded with uniforms to be sent out to the front, rather
than those of the fallen being sent back after battle. There were
plenty of other reminders of conflict and loss in the public spaces
local to her that year. Nearby Montparnasse Cemetery was dense
with graves: of the victims of the Franco-Prussian War, and the
Siege of Paris and the Commune, adults and children. At the
end of rue Froidevaux are the public gardens and the entrance
to the catacombs.

It's in these dark spaces underneath Paris that 'Madame M'
finds an organising image. It all falls into place. The children
haven't died: they have been stolen and hidden in cellars. The
soldiers who were supposedly lost in the war aren't dead either.
They're trapped in the tunnels beneath the city. The uniforms
on the lorries by the side of the road belong to these soldiers. She
can now see a coherent alternative world with its own strange
logic, one that can contain and explain confusing personal circumstances or untethered grief. The plot extends into the marital
home: any dissatisfaction or alienation is explained by the fact that
her identity has been stolen and her husband is not her husband:
they have both been replaced by substitutes.

Something else is going on, out of sight, in 1918. With uncanny
synchronicity given 'Madame M's belief in doubles and substitutions, another extraordinary feat of engineering is underway,
just north of Paris. The works digging out the *Métro* she can't
avoid, but this operation is bigger and crazier than anything she
could have imagined. 'Madame M' would have known nothing
about it – almost no one did. It was a closely guarded military
operation. Maps discovered by *Le Figaro* in the city archives

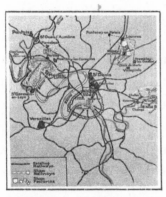

INCLUDING A PLAN FOR A SHAM PARIS IN THE FOREST OF ST. GERMAIN : THE GENERAL SCHEME (PARTLY CARRIED OUT) OF FALSE NOCTURNAL OBJECTIVES. Three zones of false objectives were planned, and one (A 2) was actually carried out. The others became unnecessary through the ending of the war. The letters on the above map show : A 1, the real district of St. Denis ; A 2, the sham St. Denis, actually created between Orme de Merlu and Louvres ; B 1, the real Paris ; B 2, the sham Paris which was to have been created between Maisons-Laffitte and Conflans ; C, a false factory district planned around Vaires.

SHOWING WHERE FALSE RAILWAYS, STATIONS, AND STREETS WERE TO BE ARRANGED : DETAILS OF THE SHAM PARIS NEAR MAISONS - LAFFITTE.

Plans for a 'fake Paris' designed to confuse enemy raiders
during the war – reprinted in the *Illustrated London News*, 1920.

show a 'Second Paris' was under construction, complete with a replica Champs-Élysées and Gare Montparnasse. The plan was devised by the Air Defence Group in the era before radar to fool German bombers, and save the real city with a decoy: a life-size fake. The *Illustrated London News* of 6 November 1920 picked up the extraordinary story after the war and printed sketches of the original maps from October 1918: 'Objectif A' and 'Objectif B'.[11]

The civic doppelgänger was to be sited in the Forest of Saint-Germain-en-Laye, in the little village of Maisons-Laffitte, on a section of the Seine which curved in a similar way to the one which ran through Paris and might fool a pilot from the air. Plans were executed by the electrical engineer Ferdinand Jacopozzi, the man responsible for the wonder of the Eiffel Tower, to illuminate 'Paris' with electric lights, lending a lifelike glow to the double, with lanterns flashing red and white on cue to pass for train lines and working points. The Armistice was signed before the plan could be finished or trialled, but life-size landmarks were mocked up in wood and canvas, including a model Gare de l'Est and a dummy suburb of Saint-Denis was completed, situated to the north-east at Villepinte. The small section that was built was quickly dismantled and forgotten. 'Madame M' didn't have any idea that this giant substitute was being put together just a few miles away from where she lived. What would she have made of the top-secret maps laying out the landmarks' doubles had she known of their existence? How might she have incorporated the project into her delusion? Could it have wrong-footed her and somehow snapped her out of it? The fake Paris project feels as incredible as one of her conspiracies. The decoy city requires us to believe secret operatives are going to the most extraordinary lengths to fool others. In that sense it would fit right into her scheme.

So how did Joseph Capgras interpret 'Madame M's false beliefs, and what explanations did he offer?

Capgras understood that there is no better way to understand delusions than to listen to the people experiencing them.

The 1923 'Madame M' case study was the legacy of over a hundred years of fine medical description. This literature first came out of France after the Revolution when physicians began more active listening and meticulous note-taking. The approach reflected a more receptive attitude and began to build a literature around the new discipline of psychiatry in its modern sense. Cases of delusion were a particularly intriguing element of this literature and the first part of the nineteenth century was a golden age of investigation into them. It was a new subject and open for debate, and leading physicians like Philippe Pinel and then his protégé Jean-Étienne-Dominique Esquirol generated lots of new thinking. Most radically, that they might be understandable from a psychological point of view; the product of trauma, rather than just a 'melancholic' imbalance of humours or demonic possession. A century passed before Capgras described 'Madame M'. Interest had waned in the meantime, but his notes pick up on these older insights about mental disturbances and repeat them for a new generation. If people were to be treated, not merely incarcerated, then a more humane approach was required, and that would demand careful listening and the recording of individual stories and symptoms, however subjective they might be.

In their 1923 paper, Capgras and his co-author Jean Reboul-Lachaux present the 'Delusion of Doubles' as a psychiatric disorder associated with women, as with other symptoms of 'hysteria'. They identify an 'imaginative hypertension' and 'psychic excitement' in their subject that leads to 'fantastic' storytelling, with a 'hallucinatory' dimension to how they interpret the world. The doctors have seen this kind of recognition disorder before, though. What interests them about 'Madame M' and makes them believe that they are describing something new and important – the reason for the great fanfare around the case study – is the particular way in which 'Madame M's recognition malfunctions: she fails to recognise only the people close to her.

Mme de Rio-Branco hardly ever notices doubles of passers-by that she comes across, and it is rare that she discovers suggestive resemblances amongst strangers; more often than not it is one and the same person who changes successively into the first double, the second, the third, etc., at intervals of a few

hours, a few days, or a few weeks... In order to explain the
triggering of the phenomenon... In any recognition, there
exists, more or less, a struggle between two emotional elements
of sensory, or memory images: the feeling of both familiarity
and strangeness... The feeling of strangeness develops in her,
therefore [...] and it jostles with the feeling of familiarity that
is inherent in all recognition. [...] With her, the delusion of
doubles is not really therefore a sensory delusion, but rather
the conclusion of an emotional judgement.[12]

Her failure to recognise people is based in her emotional state.
Capgras suggests that her 'emotional judgement' rationalises a
jumbled-up sense of familiarity and strangeness by not recognising
a person as themselves. He also points out 'Madame M's natural
predisposition to paranoia, as well as an obsessive attention to
detail. The spark – she suddenly conceives of an abduction plot –
is thrown on top of this and ignites a delusion which spreads into
other aspects of her life.

A year after the publication of the first paper on the 'Delusion
of Doubles' Capgras adapted his interpretation and described a
second case of the delusion more closely in line with the terms
of psychoanalysis, as pioneered by Freud, whose ideas were in
their ascendancy.[13] By 1929 a delusion featuring substitutes was
more often than not viewed through a Freudian lens – that is, as
a defence against internal conflicts. This view paves the way for
Leon Festinger's idea of 'cognitive dissonance' and how hard we
will try to make our clashing thoughts harmonise. According to
Freud, delusional patients develop ambivalent feelings towards
certain people close to them, and resolve the accompanying

uncomfortable, even unbearable, feelings of guilt about this by 'splitting' them into two: the 'true' individual, and the doppel-gänger, or double, onto whom they heap all the negative freight. It is easier to live with a 'double', or the idea that your loved one has been substituted, than admit to feeling a 'strangeness' towards a member of your own family, or, even worse, to the possibility that you cannot truly know the people closest to you, or they you. Capgras and Reboul-Lachaux suggested that the doppelgänger was a product of the *logique des émotions*. If a loved one no longer inspires the same feelings one has come to expect, then it cannot be that person, and must instead be someone who looks exactly like that person.[14] Freudian readings of doubles delusions also blamed unconscious Oedipal feelings – unacceptable sexual desires – for the splitting into two.

After Capgras's lecture in which he showed off his 'Madame M' and her 'unique' disorder, one of the audience came forward with a further example of the delusion. Then others from the wider psychiatric community began to rifle through old notes and speak up. The delusion turned out not to have been as rare as had first been assumed, nor was it the preserve of 'hysterical' women. In 1866 the German psychiatrist Karl Ludwig Kahlbaum had described a patient who reported imposters that were the spitting image of friends and family.[15] The delusion had been in their midst for years; they just hadn't thought to look for it.

The delusion was first referred to as 'Capgras syndrome' by psychiatrists Dupuoy and Montassut in their 1924 paper (the fourth formally reported case), but it wasn't until 1936 that a man was formally diagnosed under that umbrella. The man believed his parents were substituted doubles. There were other cases sitting quietly on the shelves, waiting to be found. In 1908, well before Capgras put pen to paper on the subject, American author and mental health reformer Clifford Beers described a psychosis he suffered from in which he believed his brother and parents were doppelgängers. Beers had even devised a 'ruse' to trick himself out of the belief, writing a letter to his brother whose identity would be authenticated should he be in possession of the letter when the brothers met. With proof confirmed, Beers claims to have successfully challenged the beliefs underpinning his delusion.

Capgras syndrome is still considered rare, but what cases there are in the literature do show a range of reaction to the substitute doubles.[16] One subject whose husband had been 'replaced' would pour tea for the doppelgänger, plus an extra cup for her husband, in case he returned.[17] One woman had such a rapport with her daughter's doppelgänger that she let the situation be and rubbed along perfectly happily with her. An air of menace and danger is never far away, though, because the question remains when a substitute double arrives: what has happened to the 'real' person? The perceived substitute is in danger, the target for all the accusations, even in an extreme case blamed for murder or abduction. The possibility of reprisal lurks around the corner.

Today, Capgras syndrome is one of the delusions most directly associated with neurological disease. Capgras himself understood that organic brain disease could exist alongside the psychological dimension when a delusion develops. He investigated delusion of doubles in old age and how it related to the degenerative processes of ageing in the brain. But it was not until the 1980s that diagnostic technology homed in more closely on the brain lesions which so often co-existed with delusions but were not originally thought to be related. Capgras syndrome is seen in dementia with Lewy bodies, and in Parkinson's disease. After the publication of Capgras's paper in 1923 old case notes came to public attention which mentioned syphilis, and there is no doubt that neurological damage has always been in the mix as a cause. It's possible typhoid fever was a contributory factor, as it may have been for Charles VI of France and his glass delusion. Organic disease doesn't seem to explain 'Madame M', but, of course, she was never scanned for signs of it. Capgras did look, as far as the technology of the day would allow, and writes that 'the physical and neurological examination doesn't reveal anything other than a slight bilateral patellar hyper-reflectivity' – abnormal knee reflexes. Regardless, we have to acknowledge that 'Madame M's symptoms remain to some extent ambiguous.

Capgras transferred from Maison Blanche to the Sainte-Anne asylum, where he spent the rest of his career, his work stalled by the beginning of the *next* world war. Sainte-Anne had been a progressive institution long before his arrival. While Capgras was preparing his landmark paper for publication back in 1922, it opened the first voluntary mental health service, meaning that patients could turn up for treatment without being detained under the old law. The hospital was a good fit for him, with his instinct to challenge dogma wherever he found it. He was remembered as being well liked by his patients: 'I have spoken to many who still remember his kindness and devotion,' writes Jacques Postel, the chief psychiatrist at Sainte-Anne during the 1980s.[18] Capgras was also remembered as being unusually open to the ideas of his younger colleagues, even if they did not share his views.

Joseph Capgras died in 1950 aged seventy-seven. He was

cared for in his final years by his nephews, Xavier and Paul Abely. Abely, also a psychiatrist, had dedicated himself, under his uncle's influence, to theories around the 'mirror sign' and schizophrenia. This was Abely's observation that people with schizophrenia spent more prolonged periods looking in the mirror than the general population. According to Abely, this extended mirror-gazing demonstrated disordered self-recognition and feelings of alienation. We can already see an afterlife here for Capgras's work into familiarity and strangeness, 'self' and 'other' in a 'Delusion of Doubles'. When Abely mentioned his uncle, 'it was in terms of his knowledge, clinical ability, and his total rejection of the grandiose clinical systems in which so many well-meaning clinical psychiatrists have become bogged down with a perverse delight that finds its equivalent only in the acedia of the Divine Comedy'.

Every clinician comes across a case early in their career that colours their research and thinking from then on. 'Madame M' was always there in the background, a life in parallel with his, a woman who had lived through the same war, but, unlike her doctor, no obituaries were written about her. She remained a source of fascination for Capgras. He returned to her story again and again over many years, thinking in new ways about it.

Whatever predispositions or diseases might have co-existed, the psychological dimension to 'Madame M's delusion is as compelling now as it was to Capgras.

'My signature has value,' she says. She wants her story to be recorded, she wants paperwork. We can hear a loud assertion of identity, of significance, of uniqueness.

Did 'Madame M' get what she wanted? The divorce she petitioned for, for example? Her belief her husband had been murdered and replaced suggests a certain ambivalence towards him at the very least. Divorce for French women had only fairly recently been re-established, in 1884 under the Third Republic. Divorce had first been introduced after the Revolution but continued to be seen as revolutionary and anti-Catholic and its social effects were feared by the aristocracy who didn't want a return to all that, and certainly not to women being allowed to petition on equal grounds to men, as they had briefly been able to. Divorce was restricted under Napoleon and banned altogether under the monarchy and only returned after the Franco-Prussian War after decades of legal fighting. At any rate, as a patient in an asylum she would not have the autonomy to set a divorce in motion. We certainly register in her voice a clear warning that things are not all right, and she did attract the attention of Joseph Capgras, for a time, a man who wanted to listen and was sensitive to trauma. To that extent the delusion was successful. 'Madame M' is clear that the authorities must act, to restore justice and order. This was a sentiment no doubt shared by many at the end of the war. Women had been enthusiastically mobilised in France, but demobilisation had been fast and without ceremony. War did not lead to the enlargement of their civic or civil rights. Unlike their European neighbours in Britain, Germany, Austria and Holland, French women did not obtain the right to vote. The civil code granting wider rights to women would not be reformed until 1938, and then only in a limited way. The rationale for this was partially economic, partly to restore the domestic structures for the returning soldiers who wanted to find their women 'as they had left them, where they had left them'.[19] Did she ever consider the conspiracy foiled? What became of her surviving child after

Capgras stopped chronicling her story? Did her husband, if he remained legally her husband, ever visit her again? Did Capgras? The records show only that she was moved to the incurable section of the asylum. They don't reveal if she ever left Maison Blanche.

Like fake Paris, 'Madame M's delusion is an extraordinary feat of imagination. The bespoke reality she makes for herself works hard for her, containing a messy life in one giant narrative. It organises her enemy – they are identity-stealers who must be stopped. It gives her a job – fighting a specific crime. Her world is well ordered and moral and it won't entertain any creeping nihilism or anarchy. Reality becomes, if not more bearable, at least more avoidable – she is fully occupied with the mental heavy lifting of keeping the delusion going. The pace and scope of the doubling is dizzying, and the world teems with impersonators, both of the people she says are dead (but who we know to be alive) like her husband, and also of the people she acknowledges to be alive, like the surviving daughter, who she sees as a series of fakes. Then there are the impersonators of 'Madame M' herself. Her story has some of the overblown magic of a magic lantern show which obscures the formative incidents and influences of her life behind the scenes. Despite a few suggestive clues, the background remains sketchy.

The flipside of the doubles, though, are the abductions and disappearings. She flags up these absences, of children and soldiers, and in doing that articulates something quietly poignant. In her alternative reality, these missing people are *still alive*. She campaigns frantically for their rescue but it's easier to believe that loved ones have been hidden, even that a substitute has been buried in place of a child, than live with the random cruelty of their deaths.

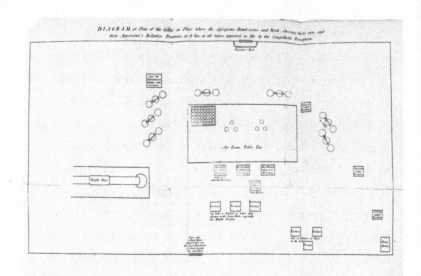

Detail from an illustration by James Tilly Matthews
for *Illustrations of Madness* by John Haslam (1810).

CHAPTER II

A Paranoid Conspiracy: James Tilly Matthews and the 'Air Loom Gang'

We move back in time by some 130 years. The date is 6 September 1793, and we're travelling north, from the Left Bank and Montparnasse, across the Seine, to Montmartre in the 9th arrondissement. The Hôtel de Biéville sits on the rue de la Grange-Batelière: an imposing neoclassical building, its vast shuttered windows topped with stone swags framing the monogram 'DB' for the Duc de Biéville who commissioned it. Until recently it was occupied by a captain in the king's personal guard, but now the French National Guard has jurisdiction. At the top of the sweeping staircase is a room guarded by gendarmes; inside sits a young man who looks to be in his twenties, head hanging down and clasped in both hands. Before all this, he was a London tea-broker. The Terror commenced by decree the previous day, enforcing the harshest measures against those suspected of being enemies of the Revolution, and the Committee of Public Safety has named this individual, who calls himself James Tilly Matthews, a spy. Rooms above and below are occupied by many others now incarcerated, with countless more in jails across the city, none of them with any prospect of trial. Civil war has spread from the Vendée and hostile armies surround France. The different faces of the Revolution, the more moderate Girondists, and the Jacobins, are turning on each other. The revolutionary government is in no mood to give special favours, least of all to a man who says he is a volunteer envoy from the British government. Matthews has had his passport confiscated, and he doesn't speak a word of French. He waits, planning his next move. 'It is not easy to find out the truth about this person,' writes the author of an occasional column called 'Revolutionary Mondays' in the newspaper *La Justice* which, a century later, looks back at the cause célèbre of Matthews and his appearance in Paris, 'the antecedents, the true conditions of this half-masked, mysterious negotiator. We only know of him what he wants us to learn…he says that he is not English, but Welsh, and that since Caesar, the inhabitants of Wales are known for their love of liberty.'

Matthews's attempt to explain himself is quoted extensively in the same newspaper:

> I may also add that my mother is from a French family of the name Tilly, who were obliged to leave France after the revocation of the Edict of Nantes. I was baptised under this name. I appear in the baptismal registers under the name of James Tilly, and, like the coat of arms in use in my country, I have always had French arms on my badge/insignia. I don't usually use the name Tilly as in my country, to use three names would be ostentatious. And so, I am happy with that of my father, except when it is some act of property, when I never omit to insert them all. The constituent Assembly has reintegrated into their rights as citizens all the descendants of French refugees. I too resolved to leave England, because I didn't like the government's principles. It is in consequence

of this resolution that I started to take a strong interest in French affairs.[1]

He adds that he does not detest the English, but their government.

This young man is falling over himself to say the right thing about his true allegiance and save his own skin. He is twenty-seven years old and in big trouble. What sort of figure does he cut? Is he still sporting the standard attire of a diplomat: silky knee-breeches, lace cuffs, ruffles, frills and a frock coat, or has this 'gadfly', as the 'Revolutionary Mondays' columnist calls him, found plainer workers' clothes in a nod to the more militant supporters of the Revolution who have seized a brief moment of power during the Terror? Matthews's primary contacts in Paris have been the more moderate statesmen of the Revolution, men like Georges Danton and Charles-François Lebrun. If he's dressed like his associates, as a gentleman, then, like 'Madame M' at the police station, he's visibly a man out of time, styled according to the diplomatic codes of an age which have gone out of date overnight. Danton and Lebrun have also just been named as suspects.

As a self-styled negotiator Matthews has been trying to keep the peace between warring factions who do not want it kept. He's persisted for several years with these intractable negotiations, demands ratcheting up on all sides, concessions denied, but trust has completely broken down. Now the French are questioning his motives, as the English soon will, and his neck is, almost literally in the age of the guillotine, on the line. His life depends on choosing the right side but power can flip from one to the other in the course of a day, an hour even. French contacts suspect him of being a double agent. His deft tightrope-walking is suspicious.

His ability to adapt himself according to his audience has been a lifesaver, but now looks like evidence of guilt. He must decide.

Matthews is then held in Paris, in a series of prisons until the early part of 1796. He's in chains, in solitary confinement and in fear of random violence and his life. By then, the revolutionary government have concluded that this so-called 'emissary' is indeed a double agent, and he is rather miraculously spared the guillotine and thrown out of France on the grounds that he is a lunatic. This may well be an administrative error in all the chaos. It's just possible that someone bothered to assess Matthews's mental capacity and took a humane view of what should be done with him, but if he was very obviously unbalanced or raving it was a rare display. He's sent packing back to England. There are no carriages for him now. He apparently walked shoeless from Paris to Calais, scavenging what he could. Over the following year he will convince himself of a sinister political conspiracy at work right under everyone's noses.

Occasionally a case emerges from the archives that allows us to see what such a delusion might have meant on an existential level for the person experiencing it. James Tilly Matthews is one such case. He is considered the first fully documented case of what became known as 'paranoid schizophrenia'. Matthews's delusion features an extraordinary mind-altering machine and a shady gang of operatives – proto-Dickensian crooks with pox-scarred hags as accomplices. This gang is sending out invisible magnetic waves from their newfangled device into the heads of the Westminster elite. The covert operation is part of a far-reaching conspiracy between French Jacobin revolutionaries and the British government. Corrupt parties both sides of the Channel are in league working to upend the political power structures in London as in Paris and install revolutionaries in Westminster.

Matthews left London as a tea-broker-turned-diplomat, with friends in high places, but when he arrives back in town he is penniless, a common felon who has been unceremoniously deported. He finds his wife and child living in abject poverty. His wife has always been unidentified, a faceless presence in his story, and we don't know what happened to her while her husband was detained in France, but it's safe to say the revenue stream from the wholesale tea trade dried up. They're in basic lodgings in the south-east of the city now and must come to terms with a new situation. This is a stark reduction in circumstances, a humiliation that's often included in accounts of a time when a delusion began to form.

Britain is a confused political scene. Wild conspiracy theories have begun to germinate here about the French situation. Conservative commentators whisper that the chaos of the Revolution might have been by the design of the 'Illuminati'. This is a society of intellectuals hoping to advance their Enlightenment principles and the talk is that they have infiltrated the Freemasons network, and the French elite. Now they're posing as Jacobins and they have overthrown the monarchy and the church in the service of their own agenda.

Confined to his basic lodgings, Matthews is entertaining notions of an alternative plot, one even more sinister and wide-ranging than anything the Illuminati are suspected of. It has crossed the Channel and infiltrated Britain. He begins to organise his intelligence. Members of the government are doing secret deals to allow Jacobin revolutionaries to overthrow the British government. He will reach out to the loyal politicians whose ear he once had. Will they hear him out? He informs the relevant officials but his allegations are ignored. On 12 September, he writes a long letter to Lord Liverpool, the minister for war, briefing him on the treason and the need for urgent action. This letter also receives no response. As an intermediary and peace broker he knows all the secrets. There are bribes paid in jewels going to and fro. What are the authorities going to do about it? He even delivers evidence to Mr Pitt himself, but the prime minister rebuffs him. He is rejected. Again and again, snubbed by the establishment he was once on good terms with. He sends one final letter. He is again ignored.

Matthews gathers some papers together and sets off for Westminster.

The next event dates the dramatic first public appearance of Matthews's delusion. Friday 4 November 1796, still not long after his return from France, Matthews sits in the front row of the public gallery in the House of Commons, visibly agitated. He waits for the right moment to strike.

There is a lull as the filing clerk inserts a clause into a bill. The man stands up and raises his right hand, in which he brandishes a piece of paper. He calls down to the chamber: 'Traitor!… Treason! Treason! I come to disclose Treason to the House.' The *London Gazette* for that week splashes with the headline 'Disturbance in the Gallery' and reports Matthews's words as: 'I wish to surrender myself to the Sargant at Arms and be examined at your Bar.'

The gallery is duly cleared and the man is frogmarched to an antechamber, where he gives his name and address as James Tilly Matthews of 6 Camberwell Grove. He then fills in a few more of the blanks in his story. The papers say he's Welsh. The

London Gazette for the first week of November 1796 says something else, quoting him that his father is Welsh, but that he is from
Staffordshire.[2] No one else picks this up this qualifying detail,
apart from, logically enough, the *Staffordshire Advertiser*.

Matthews then buttonholes everyone in the room. It's about
the treason and his thankless attempts to expose it. The demonstration in the House of Commons was a last resort. He can
prove everything to the House if he is permitted the time. He
is determined 'to let the country know the fact, or perish in the
attempt', reports the *London Gazette*. He is also being followed
everywhere by spies, even a man who 'personated him' (another
substitute double on the streets). His life is in danger. People have
been employed to kill him. The House, unsurprisingly, does not
do his bidding. They decide instead that he is 'deranged in his
intellect'. The *London Gazette* says he shows 'disappointment and
regret' at being ordered to leave Parliament.

There's more. At some point before the scene in the House
of Commons, Matthews has become convinced of an extraordinary machine at work at the centre of the plot. This is a piece
of kit called the 'Air Loom', which can control minds, weaving
'airs', or gases, into a 'warp of magnetic fluid' which is then
directed at its victim. Revolutionaries with their fingers on the
buttons and levers are manipulating the thoughts of serving
British politicians.

> Near every public office an air-loom is concealed, and if the
> police were sufficiently vigilant, they might detect a set of
> retches at work near the Houses of Parliament, Admiralty,
> Treasury, &c, and there is a gang established near St Luke's

Hospital…all the persons holding high stations in the government are held impregnated. An expert of the gang, who is magnetically prepared, contrives to place himself near the person of a minister of state, also impregnated, and is thus enabled to force any particular thought into his mind and obtain his reflections on the thought so forced. – Thus, for instance, when a Secretary of War is at church, in the theatre, or sitting in his office and thinking on indifferent subjects; the expert magnetist would suddenly throw into his mind the subject of exchange of prisoners. The secretary would, perhaps, wonder how he became possessed of such a subject, as it was by no means connected with his thoughts; he would, however, turn the topic in his mind and conclude that such particular principle ought to form the basis of the negociation. The expert magnetist, having, by watching and sucking, obtained his opinion, would immediately inform the French Minister of the sentiments of the English Secretary, and by such means become enabled to baffle him in the exchange.[3]

The gang are influencing at the highest level. Even the prime minister William Pitt is no match for the magnetic forces and has become a 'mere puppet' to the villains who plan to assassinate him. Matthews has survived on his wits. This is dangerous work. He may look like a double agent, but he's actually a whistle-blower keeping tabs on the plans for 'the republicanising of Great Britain and Ireland', and 'disorganising the British navy'. Our daring hero falls over himself to declare his allegiance, but this time to the British, downplaying any affinity with the French. Cells of the Air Loom are concealed in numerous basements and cellars across London, even in the grounds of the hospital itself. We think of 'Madame M' and the Paris of more than a century later with its basements full of stolen children.

Bow Street Magistrates hear the case and Matthews's family plead with the authorities that he is sane. Matthews is a tricky proposition. Here is a delusional conspiracy theorist but some of the most implausible elements to the story he told police are demonstrably true. He was a tea-broker but newspaper reports confirm that he had been in Paris during the most tumultuous years of the Revolution and at a pivotal moment in world history. Brokering peace instead of tea is quite a career change.

He appears before Lord Kenyon at Lincoln's Inn Fields, but after petitions from his parish authorities in south-east London, who have had enough of dealing with this troublesome 'pauper' on their patch, he is committed.

On 28 January 1797, the year after he was sent packing from France, Matthews enters the Bethlehem psychiatric hospital, better known as Bethlem, or 'Bedlam'. At the time, the hospital was still situated at Moorfields, just beyond the northern wall of the City of London. Records show that he is moved to the incurable section of the hospital the following year, on 21 January 1798.

'James Tilly Matthews put on the incurables', 21 January 1798, Bethlem Hospital Incurable Patient Admission Registers.

Matthews had been declared a 'lunatic' by the French authorities, but was never admitted into the French psychiatric system
that was pioneering a more psychological, humane approach to
mental illness. He was brought instead to the most celebrated
British asylum of the day. It was a different picture in Bedlam. The
hospital Matthews arrived at was, on the exterior at least, a showy,
purpose-built affair. Designed by the famed polymath Robert
Hooke, its grand scale advertised a great benevolence towards the
predicament of lunatics. Stone pillars at the entrance cautioned
as well as impressed visitors with the figures of 'Melancholy' and
'Raving Madness' carved in Portland stone at the top. The hospital
had been built, however, on the 'town ditch' – essentially a rubbish
tip – so its foundations were unsound. By the time Matthews got
there, the place was falling down; the enlarged cells and well-lit
galleries of Hooke's design now damp and dilapidated, with
uneven floors and a leaking roof. The patients there were still
manacled, still enduring regular cold-bath treatments, bleeding
and all sorts of other violent restraints and violations. The post
of physician there had historically been awarded on a charitable
and honorary basis, and a salary was a relatively new thing for
a man like John Haslam, the doctor-at-large and apothecary
at the hospital. Most physicians were in and out, making their
money 'mad doctoring' in private practice, but as an apothecary
Haslam was on site daily. This may go some way to explaining
his rise at the institution.

The term 'Bedlam' was, even then, synonymous with chaos.
The asylum featured in the final scene of Hogarth's series of
paintings 'The Rake's Progress' in which protagonist Tom lies
almost naked on the floor, his loyal wife Sarah weeping beside

him, as an asylum tourist and her maid look on. Visitors were still coming in large numbers on high days and holidays to gawp at the spectacle, and spot its most famous residents despite the squalor of the place; perhaps even because of it. Years of war with revolutionary France had drained the public coffers and made refurbishment of institutions like this a low priority.

On 18 March 1797 a woman interrupts the morning meeting of the Bethlem Subcommittee. 'I am the wife to Mr. Matthews and demand to know by what authority my husband is detained.' Matthews's admission papers are shown to her but she won't be fobbed off. Eventually she receives an order that she is longer allowed to visit her husband. She makes his behaviour worse. She's kept a nameless, indistinct figure, on the other side of the walls, but she will haunt the hospital authorities.

Who knows exactly when the two men first set eyes on each other but it's John Haslam we have to thank for Matthews's promotion from a bizarre diplomatic anecdote in a French newspaper to a landmark case study of delusion. Haslam is in charge of Matthews's treatment at Bedlam and he compiles a book about the case: *Illustrations of Madness*. Haslam's treatment of the story is highly original. His turns a 'first person' account of a delusion into a page-turning caper. He is puppet master to his patient but at pains to say that Matthews was keen to get the story out there and shared his story willingly, collaborating with Haslam by drawing the diagrams and character sketches himself. This is the first time an entire book has been devoted to the delusional imagination of one individual case. Haslam subtitles his book 'Exhibiting a Singular Case of Insanity, and a no less remarkable difference in Medical Opinion: Developing the

Nature of Assailment, and the Manner of Working Events; with a Description of the Tortures Experienced by Bomb-Bursting, Lobster-Cracking and Lengthening the Brain'.

John Haslam lives at close quarters with Matthews for more than fifteen years before he presents the delusion to the public for the first time. Our intermediary keeps careful record of Matthews's 'insanity' over the many years of his confinement and he takes us on an up-close, street-level tour of Matthews world. The resulting guide to one man's delusion is in the eighteenth-century version of Glorious Technicolor, with tracking shots along dirty, crowded streets. The atmosphere is seedy and oppressive. The dramatis personae are rounded up for us:

> Mr M. insists that in some apartment near London Wall there is a gang of villains profoundly skilled in pneumatic chemis-try... The assailing gang consists of seven members, four of whom are men, and three women. Of these persons four are commonly resident, and two have never stirred abroad since he has been the subject of their persecution; of their general habits little is known. Occasionally they appear in the streets, and by ordinary persons would be taken to be pickpockets or private distillers. They leave home to correspond with others of their profession; hire themselves out as spies, and discover the secrets of government to the enemy, or confederate to work events of most atrocious nature. At home they lie together in promiscuous intercourse and filthy community.[4]

Haslam promises special access to this filthy community through the eyes of his swashbuckling investigator on the ground. Matthews's words are faithfully transcribed, Haslam insists, and he is 'contented and grateful' to have the plot brought to public attention. Matthews holds a lantern up to a villains' gallery, illuminating the characters from his imagination. Bill 'exerts the most unrelenting and murderous villainy', and 'has never been known to smile'; his second-in-command, 'Jack the Schoolmaster' is 'about 60 years of age. It is not well ascertained if he wears a wig, but he generally appears in the act of shoving his wig back with his forefinger, and frequently says: "So you shall, when you can ketch [catch] us at it."' The leader, 'Bill the King', is ruthless and 'surpassed the rest in skill, and in the dexterity with which he worked the machine. He is about 64 or 5 years of age, and in person resembles the late Dr De Velangin [an eccentric doctor interested in diet and nutrition who worked for a time at Bedlam] but his features are coarser; perhaps, he is a nearer likeness to the late Sir William Pulteney [the fabulously wealthy and Pitt-supporting parliamentarian] to whom he is made a duplicate.' Brief flashes of recognisable London society personalities make the scene disorientating, both familiar and unfamiliar, like faces in a hall of mirrors. The crude copies walking the streets of Matthews's London call forward to the frauds who lurk around in 'Madame M's Paris more than a century later with their obvious post-operative scars. 'Sir Archy' wears a 'drag coat' with breeches that button between the legs. 'Some of the men assert that Sir Archy is a woman dressed in men's apparel...a low-minded blackguard, always cracking obscene jokes and throwing out gibes and sarcasms. In his speech there is an affectation of a provincial accent.' He uses a magnet to manipulate 'brain sayings'. The operator of the machine is a sinister, disfigured lady known only as the 'Glove Woman': she

always wears cotton mittens as she has 'the itch'. She is 'pockfret-
ten' – scarred by pox – with 'downy hair' on her upper lip and
chin, she wears a 'common fawn-coloured Norwich gown', a plain
woven 'camblet shawl' and a 'chip hat covered with black silk…
The rest of the gang…are constantly bantering and plucking at
her like a number of rooks at a strange jack door [jackdaw]: she
has never been known to speak.' Augusta is superficially charming 'ordinarily dressed as a country tradesman's wife, in black,
without powder', but 'exceedingly spiteful and malignant' when
crossed. She roams London's West End as an undercover agent
with other gangs, her specific objective with the magnets is to
influence women. Charlotte has 'the appearance of a French
woman, being a kind of ruddy brunette'. They keep her 'nearly
naked and poorly fed. Mr Matthews is led to suppose that she
is chained; for she has sometimes stated herself to be equally a
prisoner with himself. She always speaks French, but her language
and brain-sayings are conveyed in an English idiom.' You can't
miss the confusion of French and English identity in the character of Charlotte; she embodies the confusion of Matthews's
influences and anxieties about allegiance to one side or another.
Perhaps she also reflects his own disjointed memories of his time
spent chained up during the Terror.

There was already a high-profile patient under Haslam's care
at Bedlam when Matthews came through the doors: a docile and
enigmatic woman, fond of Shakespeare and gingerbread, called
Margaret Nicholson. The pair would go on to spend many years
as neighbours. She was a mother figure to the other patients just
as he later become a father figure. Nicholson was sixteen years
his senior, and had already been at Bedlam for a decade when he

turned up carrying his fully realised world and its cast of characters. She had been admitted after attempting to stab King George
III with a blunt dessert knife and was something of a celebrity spot
for the sightseers to the hospital who would beg an audience with
her. It's conceivable that Matthews was given special permission
to cross into the women's section of the hospital and meet her.
They would hear about each other, and they had, after all, a lot
in common. Both exhibited delusions of grandeur. According to
Haslam, Matthews had occasionally been known to lose his usual
composure and throw his weight around, like the 'Napoleons'
who arrived a few years later at a number of French asylums, as
if he were 'Emperor of the whole world, issuing proclamations
to his disobedient subjects and hurling from their thrones the
usurpers of his dominions'. This line could have come straight out
of Nicholson's mouth as she demands her usurper King George
III acknowledge her claim to the throne. Matthews's delusion
of persecution puts him firmly at the centre of the universe and
assumes that all eyes are fixed on him. As with 'Madame M', he
is powerless, but this is predicated on the world caring enough to
disenfranchise him. Matthews's life after his diplomatic mission
runs parallel with Napoleon Bonaparte's rise, but with the two
men going in opposite directions. 'We told you that you were
Buonaparte's talisman', the gang gloat, 'and that we would work
him up to as high a pitch of grandeur as we would fix you below
the common level of human nature.' Haslam describes Matthews
with an arched eyebrow as 'our own resident Un-Napoleon'.

Haslam has good reason to resent Matthews. The doctor saw
sanity and insanity, like day and night, as two states that could
not, by definition, co-exist. Matthews's presentation makes it very

difficult to be so black and white about the question of whether or
not he's 'out of his mind'. He is such an accomplished draughtsman that when a competition was launched for a design for a new
Bedlam on its new site on St George's Fields in Lambeth, south
London (the building which is now the Imperial War Museum) he
submitted designs. He didn't win, but his efforts so impressed the
governors that he was paid £30. Over the years of his incarceration he was apparently a mild-mannered character. His doctor
said he could even be an 'automaton' waiting for others to take
charge of him (rather like 'Madame M's listlessness). At other
moments he is described as scholarly, peacemaking and well liked
by fellow inmates. He was allowed to garden some land among
the tumbledown, dated exterior walls which he 'brought to a high
state of cultivation'. One of his unrealised schemes in Paris had
been to solve the food shortages with a network of city gardens
and he made plans for a kitchen garden at Bedlam. His family
stood by him for the duration of his time at Bedlam, and continued to lobby for his release, although they were barred from
visiting him. Banning visitors to Bedlam was very unusual and, as
Mike Jay suggests in *The Influencing Machine*, only strengthened his
belief that he was a political prisoner and an enemy of the state.

Matthews is a real threat to Haslam's authority, but their relationship remains ambiguous. It is Haslam who gives Matthews the
pens and other materials to enable him to draw his private world
(just like wardens at the beginning of the twentieth century will
give our milliner 'Léa-Anna B' Shakespeare plays to read while
she waits for word from King George V) and he fosters Matthews's
evident talent as a draughtsman. He seems, on one level at least,
to be captivated by his patient's imaginative creation – the Air

Loom and the realm in which it operates – even as he sneers at the reality of this world in his commentary. The relationship between this doctor and patient also dramatises a very old question, the same one opposing religious authorities used to ask: namely, who gets to decide what is delusional, and what is not?

Matthews's exceptional technical skill gives his diagrams a lifelike precision. He gets down to the nitty gritty of how all these unseen forces are worked and exploited by this fantastical cast of agents. And we loop around again to the perennial and knotty question of what a delusion might *mean*.

The air from the device is a putrid cocktail of 'stinking human breath, mortification and of the plague', 'seminal fluid' and arsenic among its component parts. The machine begins a quasi-mechanical process whereby this air assails a person, causing them to act and feel in a desired way through planted thoughts, or the more torturous 'lobster-cracking' to which Haslam refers in the subtitle of the book, when the pressure from the magnetic atmosphere stops a person's circulation, as if they're caught in the teeth of tongs. Forces are also channelled through ordinary domestic appliances: 'apoplexy-working with the nutmeg grater', for example, or, most subversively of all, a cylindrical mirror held up to the target so that all 'thoughts are made to assume a grotesque interpretation; and the person assailed is surprised that his fixed and solemn opinions should take a form which compels him to distrust their identity, forcing him to laugh at the most important subjects. It can cause good sense to appear as insanity, and convert truth into a libel; distort the wisest institutions of civilized society into the practices of barbarians, and strain the bible into a jest book.' In other words, this system can turn the world upside down in a moment and make a laughing stock out of anything and anyone.

The gang work as a team. One member sucks out existing sentiments, another forces in new ones. There is bawdy comedy when bodily functions are enlisted: magnetic fluid, for example, is extracted through the anus 'bubble by bubble' to dastardly ends. Magnetism as a vehicle for psychic communication is perhaps the most extraordinary and fearsome trick. An 'inferior member of the gang (generally a novice) is...furnished with a bottle containing the magnetic fluid which issues from a valve on pressure. If the object to be assailed be sitting in a coffeehouse, the magnetic practitioner hovers about him, perhaps enters into conversation, and during such discourse, by opening the valve, sets at liberty the volatile magnetic fluid which is respired by the person intended to be assailed. So great is the attraction between the human body and this fluid, that the party becomes certainly impregnated.' The communication and control happen 'without producing the ordinary vibrations of the air', lodging in the cavity of the ear while the bystander is none the wiser. The gang even work at night, forcing their grotesque images into the intellect in a way which is both 'wonderful and distressing'. This can happen to you when you're asleep or drinking your coffee. You might get an apprentice on work experience. It's no laughing matter, though. It was this gang who persuaded the authorities to cook up false allegations that Matthews was mad in order to invalidate his claims. They got him committed. The further a person goes from the pneumatic machine the weaker its hold on him, but you have to get a thousand feet away before it loses its controlling power. The 'interposition of walls causes but a trifling difference'. The conspiracy can penetrate the hospital and reach him. They can still reach him and will eventually kill him. He can't escape, even inside fortress Bedlam, just as 'Madame M' will be trapped in Maison Blanche with doctor doubles who are involved in a plot to poison her.

To Matthews, this is no caper. This is a crime of high treason, and it's live. The situation is perilous.

Today, Matthew's ideas would be seen as 'paranoid'; a category of delusion which might include ideas of 'reference' (the belief that harmless or coincidental events relate directly to you, when they don't) and persecution (believing others are observing you and may be trying to harm you). The word 'paranoia' was coined by Hippocrates to describe the delirious ramblings people experience when they have a very high temperature. He used two Greek words – 'para' meaning beside and 'nous' meaning mind – to create a word meaning literally 'out of one's mind'. The term was picked up and eventually came to refer to a disorder of mind that did not involve delirium. The World Health Organization's classification system for psychiatric phenomena, built on work at the Maudsley Hospital in the 1960s, established the principal delusion types: grandiosity, reference, jealousy, pregnancy, catastrophe, religion, guilt and others. Over the last century paranoid delusions have become the most common of all.

With the medicalisation of mental health conditions at the end of the nineteenth century came the diagnosis of 'dementia praecox', which would soon be known as schizophrenia, whose schema swallowed up delusions as one of many symptoms. Matthews was retrospectively labelled with this diagnosis and within its framework the Air Loom delusion was 'un-understandable' in terms of psychology or life experience.

It's true that Matthews as he appears in *Illustrations of Madness* is a larger than life character, a showman demonstrating his visions whose human features are obscured behind smoke and mirrors. Can we catch a glimpse of the man? Where did the path that took him to Bedlam start?

Matthews became clearly unstable after years working as a negotiator in revolutionary France. But what made him decide to go to Paris to attempt such an impossible diplomatic task in the first place? Can we even be sure what name he went by? By his own admission he was in the habit of using different config- urations of his surnames at different times depending on which country he was in and how much of a show-off he wanted to be. Most biographies of Matthews give his birth date as 1770 and state that he was originally from Wales, and little is generally known of his early life. Welsh origins have been assumed thanks to his statement to Bow Street Magistrates where he says the name Matthews is from his Welsh father. A detail in the administrative documents from the day of his arrest in London rewards another look. He stipulates that he was baptised James Tilly, under his French Huguenot mother's name, not under Matthews. This leaves more variables for a researcher trying to find traces of him in the records, especially given the fact that small changes or discrepancies in spelling were common in registers at this time.

A throwaway remark by Matthews, supposedly from his own mouth while making his statement, provides a lead and helps focus the search. It's been sitting there all along, overlooked in the *London Gazette*'s report of Matthews's antics at the House of Commons.

His father, he says, was from Wales, but he is himself in fact a 'native of Staffordshire'. Any trumpeting of national identity by Matthews should be taken with a big pinch of salt. We know that he would say anything to stay on side. But naming the country of Staffordshire as the place where he grew up has an authentic ring, since there's no obvious reason to lie about it either way. It's the *Staffordshire Advertiser* of 12 November 1796 (quoting the *London Gazette*) that brings us the breaking news of the drama in the House of Commons – did the paper believe this to be a local man?

Modern search engines mean it's now possible to cross-reference variant spellings on a name with a possible new location in the online records.

There is a baptism record for a 'James Tilley Matthews' (Tilley with an 'e') for 8 March 1766 in St Peter's Church, Kinver (or Kinfare), Staffordshire, England. The entry goes on to call him 'son of Edward Matthews Jane his wife'.

Parish baptisms, 1766, Kinver Staffordshire: 'James Tilley son of Edward Matthews & Jane his wife, bap. March 5th.'

According to another Parish register, this time in London, a 'James Tilley Matthews' (again, Tilley with an 'e'), aged twenty-one, married Elizabeth Sarah Gibbs, also aged twenty-one years, on 20 April 1787 in the bride's parish of St George's, Bloomsbury, Camden.

The marriage record of James Tilley Matthews and Elizabeth Sarah Gibbs, St George's, Bloomsbury, 20 April 1787.

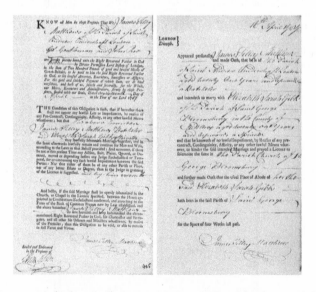

Marriage bond and allegation, James Tilley Matthews of the Parish of St Andrew Undershaft and Elizabeth Sarah Gibbs of the Parish of St George's, Bloomsbury, 16 April 1787.

Four days earlier, on 16 April, there is as marriage bond recorded – a sworn statement that you are free to marry on pain of a substantial fine, in this case of £200 – in which a 'James Tilley Matthews', 'a bachelor', is described as being 'of the parish of St Andrew Undershaft'.

Can we connect the two men? We have a very few clues as to what our Matthews was up to before his adventures in France began but he's clear about his occupation pre-spying. He was a 'tea-broker' and he gives his address as 84 Leadenhall Street, in the City of London.

This James Tilley Matthews's parish when he married is St Andrew Undershaft, the church which sits just off Leadenhall Street on St Mary Axe in the city of London (it's still there, a rare example of a building that survived both the Great Fire and the Blitz). It is less than a three-minute walk from St Andrew Undershaft to 84 Leadenhall Street, where our Matthews says he worked as a tea broker, before his diversion into French/English diplomacy. St Andrew Undershaft is, in fact, the *closest* church to 84 Leadenhall. The James Tilley Matthews who was born in Kinver in 1766 would be the right age in April 1787, just turned twenty-one, for the two records to match. If it's our man, his long-time unidentified wife now has a name, Elizabeth Sarah, and family in Bloomsbury. After so many years as a nameless presence to the side of her husband's story, she starts to take on some definition. The area surrounding the impressive Hawksmoor church where they married was developing fast into a well-to-do area, a world away from the slums of St Giles down the road, with elegant garden squares and its centrepiece, the British Museum.

If our James Tilly Matthews was born in south-west Staffordshire in 1766, then the Industrial Revolution was all around him as he was growing up. By the late eighteenth century five 'slitting mills' were working in the county which includes Kinver, more than anywhere else in Britain at the time. The mills would slit iron bars into rods to be made into nails in the surrounding Black Country. Staffordshire was at the forefront of the charge for innovation in industrial processes, in pottery, iron, coal and glass. The scale of the mechanisation would have meant a multi-sensory assault on any local, particularly on a child, the sights and sounds both awesome and troubling as the process accelerated to an unrecognisable future society.

He may well have come to London to serve an apprenticeship, as young men from almost all tiers of society did at this time, looking to qualify in a line of work which offered good prospects. If a father had contacts in the tea business – this was usually how positions were secured – indenture fees were often paid for a son's apprenticeship. Tea was a booming business with a bright future in the 1780s and 1790s, the drink now exceptionally popular with the middle classes, and drew many young people into the City of London.

East India House was the focal point of Leadenhall Street in the eighteenth century as the headquarters of the East India Company, which held the monopoly for the import of tea from India and China. Tea auctions were hosted here, placing London at the centre of the international tea trade at the time Matthews was brokering the stuff. These auctions of imported tea were held quarterly and the commodity traded 'by the candle', meaning that a candle was lit with each new lot and, when an inch had burned away, the hammer would fall, keeping proceedings moving at pace.

Armed with the good prospects of a career in tea, Matthews marries Elizabeth Sarah just a month after turning twenty-one, also the age at which an apprentice was allowed to make his living in the City of London.

Five years pass before he first travels to France. Presumably he was working during this time for, or through, whichever importers or dealers in tea had premises at 84 Leadenhall (the company name, like the building itself, is now lost). Becoming a Freeman of a livery company like the Drapers, who often took tea dealers and brokers into their company, could give a man a more secure footing. Regardless, there was every chance that, in the early 1790s, he was doing rather well. Elizabeth and James started a family.

Matthews's earlier education is harder to make sense of. The penmanship he later demonstrated at Bedlam was admired even by the doctor who loathed him for most things – he also drew frontispieces for some of Haslam's special volumes, and the fonts and designs that he set were apparently indistinguishable in quality from those of a publishing house. He may have learned callig-raphy while in Bedlam. We know his persecutor-turned-mentor was in the habit of providing him with whatever equipment he needed.

At some point, going about his daily business in London, Matthews came into the orbit of David Williams, a Welsh cler-gyman and prominent religious and educational reformer. This contact would change the direction of his life. If we take him at his word, that Matthews's father was indeed Welsh like Williams, perhaps there was a connection between the two senior figures. We know the younger Matthews became acquainted with David Williams because it was he who led the trip to Paris in late 1792 that Matthews joined when he made his very first venture to France. The expedition also included Joseph Priestley, the renowned experimental scientist who is best known for his work on the chemistry of gases. Williams was the magnetic centre of a high-profile discussion group which drew in scientists and other public figures of the day. He was a one-time associate of Benjamin Franklin and regularly hosted events in a chapel in Margaret Street in central London. These sessions tackled head-on the pressing issues of the day such as governance, and constitutional issues facing prospective new republics.

How did Matthews find himself among this heavyweight intellectual company? An obvious entrée opportunity were the public lectures run by Williams. There's no evidence of an extensive schooling in Staffordshire, or of family money, and Matthews has many of the hallmarks of an autodidact with natural ability, acquiring a high level of skill in a variety of disciplines. When he is interviewed after his arrest in London, he claims to have left England initially because he 'didn't like the government's principles'. An ambitious young man with his sights set on becoming a person of influence – and doing well enough in business to pay for it – could take private tuition for the purposes of self-improvement. Williams was known to adopt private pupils of 'defective' education who wished to take part in public life.

Williams's circle is a natural draw for Matthews. The members are all renowned and idealistic republicans keen to roll up their sleeves and get to work actioning the project in France. Having fallen in with the group, Matthews receives an offer that would turn the most level of heads: an invitation to Paris – to the very centre of the political action in Europe. They are proposing to help to draft a French constitution. It's a grand undertaking. Matthews steps up and joins the deputation, assured of the opportunity and adventure of a lifetime. On the boat to France, Williams and Priestley and the others had a lot to talk about and agree upon. Did Priestley mention anything on the way over about exciting discoveries in the field of pneumatic chemistry – about gases even? Did anything of the political and scientific zeal of these men rub off on their young companion? If the James Tilley Matthews from Kinver is our man, that makes him twenty-seven when he joins this trip to Paris, a crucial few years older than some biographies have him. All the evidence points to a curious and tenacious man who couldn't wait to dive into the wider world and its affairs. He is a natural conduit for the wide-eyed conviction of the age, but he soon acquires a preternatural level of confidence. He makes several trips to France, flitting between British parliamentarians and their French counterparts, even troubling Prime Minister William Pitt himself. Looked at through a modern lens, we might see early signs of the delusions of grandeur that can characterise schizophrenia. He gives himself a job – as an agent of peace between nations.

The central argument of Haslam's book about the case pivoted around the question of whether or not James Tilly Matthews was evidentially sane or insane. The story has been passed down the generations as a crazed *Boy's Own* adventure of the Georgian era. Haslam's agenda gets in the way of a clear view of Matthews. Can we see behind the artifice to what was going on around him in the real world; what was influencing him, by the way, as he went about his business on the streets of Paris and London before his arrest?

First, we might ask where the Air Loom, an extraordinary feat of imaginative engineering, sprang from? Where did the idea set seed?

Illustrations of Madness mentions a formative encounter. While he was locked up for spying in Paris, killing time with another young buck who had found himself on the wrong side of the regime, it seems the conversation came round to new scientific thinking. This was way beyond anything he would have heard from Joseph Priestley. It was second-hand gossip about the cutting edge of physics and chemistry and how minds might be able to talk to each other through connections that weren't visible to the naked eye. It was all new to Matthews, but he was open to it:

> The first hint Mr M received of the possibility of such sym-
> pathetic communication was in France, before the period
> of his confinement. He there, in one of the prisons, became
> acquainted with a Mr. Chavanay, whose father had been cook
> to Lord Lonsdale. One day, when they were sitting together,
> Mr. Chavanay said, 'Mr Matthews, are you acquainted with
> the art of talking with your brains?' Mr. M. replied in the
> negative. Mr C. said, 'It is effected by means of the magnet.'

And just like that, apparently, Matthews was introduced to the reality of invisible forces and their power to change minds. A chat around the subject would have led inevitably to an honourable mention of a certain Viennese Dr Mesmer, who in recent years had been expounding a new 'hypnotic' theory to the bourgeoi-sie of Paris who were in thrall to the idea. 'Mesmerism' was a new conception of natural forces organised around the idea of an invisible substance that passed through the air, an 'animal magnetism', that was present inside people's bodies and which a doctor could manipulate to various ends.

Demonstrations of mesmerism were banned during the Revolution, but when Matthews made his first few visits to Paris during the early 1790s, although Mesmer himself had already quit the city, these demonstrations were still commonplace, and members of the public could elect to be in the audience, or even participate. Matthews was too late to see the first stage for these demos, in a set of rooms at the luxurious Hôtel Bullion, on the rue Coq-Héron in the 1st arrondissement, where Mesmer set up in his practice shortly after arriving in Paris in January 1778. Mesmer had been derided and discredited in Vienna, but in Paris he found himself with up to twenty patients a day and, with his sessions overrun on a one-to-one basis, he had to up his game. A New England society magazine describes the atmosphere of Mesmer's rooms with breathless wonder. This was an immersive experience:

> In all Paris there were no apartments so charmingly furnished as Monsieur Mesmer's. Richly-stained glass shed a dim, religious light on his spacious saloons which were almost covered with mirrors; orange blossoms scented all the air of his corridors; incense of the most expensive kind burned in antique vases on his chimney pieces; Aeolian harps sighed melodious music from distant chambers...[5]

Mesmer had devised a piece of apparatus called the 'baquet' which allowed him to treat more people at a time in a version of group therapy. This baquet was a large wooden tub containing sixteen bottles in a radial pattern. Each of these bottles was filled with iron filings, powdered sulphur and crushed glass. Iron rods projected from the vessel with powerful magnets at their base, and participants would hold hands, allowing the animal magnetism to flow between them, hoping to experience its healing effects.

An English doctor who observed Mesmer at work describes in more detail how the baquet was used in the group therapy:

> In the middle of the room is placed a vessel of about a foot and a half high… It is so large that twenty people can easily sit round it: near the edge of the lid which covers it there are holes pierced corresponding to the number of persons who are to surround it; into these holes are introduced iron rods, bent at right angles outwards, and of different heights, so as to answer to the part of the body to which they are to be applied. Besides these rods, there is a rope which communicates between the baquet and one of the patients, and from him is carried to another, and so on the whole round. The most sensible effects are produced on the approach of Mesmer, who is said to convey the fluid by certain motions of his hands or eyes, without touching the person. I have talked with several who have witnessed these effects, who have convulsions occasioned and removed by a movement of the hand…[6]

Apparently, Mesmer would often conclude his treatments by playing some music on a glass armonica, and the overall effect would be hypnotic. The cultivated, fragrant and spooky atmosphere of these gatherings feels more like a seance than a medical procedure. Mesmer's salons, like the Phantasmagoria, are operating well before the spiritualist movement entered the stage. Spiritualism and the seance of the popular imagination didn't materialise until 1848, in a bedroom in a clapboard house in upstate New York where the Fox sisters, eleven-year-old Margaretta and fourteen-year-old Catherine, lived. The sisters claimed to be communicating with the dead via 'raps' on wood and crowds began to assemble around a table in the Fox home, hoping to hear these 'raps' which the sisters passed off as messages from the deceased. With their fame spreading, Leah, a third Fox sister and in her thirties, decided to take the family on tour, first of the US and then overseas, to appear before spellbound audiences. Thus a homespun cottage industry became an international movement. Mesmer's salons pre-dated the Fox sisters' roadshow by many decades but his ideas drifted across the Atlantic and his influence was felt in this craze for the occult that emerged in the 1840s. His method now seems prescient of many of the religious practices associated with spiritualism.

In Paris in the late eighteenth century the public were already showing an appetite for evidence of unseen powers. Mesmer's audience could choose from a variety of theatrical demonstrations around town, whether that be the healing powers of medical electricity, or the thrill dread of the Phantasmagoria and the magic lantern which presented ghostly apparitions, doppelgängers and the resurrection of the dead.

Marie Antoinette was a devoted follower, but an enquiry led by scientists declared Mesmer a charlatan and he fled the city in 1785. To others, however, he remained a genius. By the time Matthews arrived in Paris Mesmer was long gone, but the waves he generated were still moving around the city. Over the course of their conversation in the cell, Mr Chavanay might well have mentioned Mesmer's protégé, Antoine Hyacinthe de Chastenet, and, more significantly, Chastenet's brother, the Marquis de Puységur, who continued to draw followers while Mesmer was in exile in Switzerland. The institute run by Puységur in Strasbourg was disbanded by order of the revolutionary government and Puységur paid his dues with a few years in prison, but mesmerism and magnetism continued to exert influence for more than a hundred years. Its appeal reached all the way to the ladies' magazines of the 1920s where practitioners advertised to women like 'Madame M' and her clients. Although Matthews did not meet the man himself, Mesmer exerted a magnetic pull on his psyche, as he did on so many. Looking at the last remaining example of a Mesmer baquet, preserved in the Musée d'Histoire de la Médecine et de la Pharmacie in Lyons (acquired in 1846 by a Bavarian furrier who stumbled across it in a house clearance) it is hard to miss the references in the Air Loom and the similarities are suggestive that Matthews had seen one, or at least had one described for him in great detail.

Mesmer's centrepiece was a prototype for the Air Loom. The Heath Robinson contraption is re-versioned in Matthews's imagination. On the face of it both objects are crude, almost naive examples of industrial engineering, but both also cast forward to a world where the most powerful networks are functioning out of sight in the ether. Both objects are inventions straight out of a 'Steampunk' science fiction fantasy, but arrive nearly two hundred

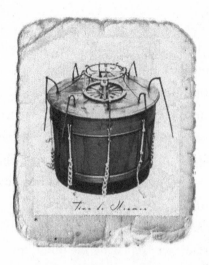

A surviving example of Mesmer's 'baquet' at the Musée
d'Histoire de la Médecine et de la Pharmacie, Lyon, France.

years before the genre was conceived. Steampunk presents an
alternative Victorian Times, hissing and clanking with a combination of retro and futuristic steam-powered machinery. The Air
Loom is Georgian, but plays by the same rules: the device boasts
fantastical powers, but it quotes real-world, analogue technology.
In *Illustrations of Madness*, the gang 'proudly boast of their contempt for the science of the present era'. Matthews's Air Loom
agents laugh at redundant old-world technologies. The machine
he conjures up is cutting edge but designed to bring about harm.
It animates an unease about the rate of scientific discovery; how
each new insight into natural laws could be used, who would be
in the know and who left behind, how it could disrupt day-to-day
life, where it all might lead.

This was at a time when a great number of invisible forces
were being discovered and utilised everywhere – the force of

gravity, for example; electricity, even magnetism itself – and scientists were beginning to identify and play with them. The realm of the supernatural was now enlightened science. In his writing about the future, Arthur C. Clarke included three 'laws'. The third law proclaimed: 'Any sufficiently advanced technology is indistinguishable from magic.'[7] The line between vanguard science and charlatanism was becoming more difficult to draw with confidence – pneumatic chemistry was a controversial field. A certain amount of magical thinking was required in order to comprehend each technological advance anyway. These newly identified energies were replacing the other 'invisible forces' in which countless generations had believed – forces like those which had accounted for madness: God's revenge on sinners, for example, or the powers of the devil and demonic possession. Once you believed in mesmerism it wasn't such a great stretch for someone like Matthews to translate elusive but powerful forces of nature into a set of notions which explained why he was so mentally unbalanced. Currents of magnetic fluid flowing between people are a neat explanation for any Romantic or existential anxieties about who is in charge now. The currents move through any wall or person in Matthews's delusion. 'Léa-Anna B' will come up with something similar in the 1920s as part of her delusion of erotomania. The force she identifies is an almost ectoplasmic cloud she calls 'La Morve' which mingles among people and influences her actions.

When Matthews visited Paris in the early 1790s, a massive civic project was underway beneath the streets. This was the grisly process of moving bones from the overwhelmed cemeteries to the derelict quarries just beyond the city. By day, Matthews went about his business in the city meeting politicians, building his diplomatic reputation, and by night labourers stocked the catacombs with the remains of Paris's dead. The ancient stone mines in the 14th arrondissement had caved in back in 1774, swallowing up three hundred metres of rue Denfert Rochereau, in the area that would become the Montparnasse of 'Madame M'. A lack of space for the dead, on top of poor city sanitation, incubated disease and an intolerable stench. Louis XVI conceived an idea for solving this crisis and the process of disinterring bones from their graves began in 1787, away from the eyes of Parisians who were deeply unhappy at breaking into the final resting places of their dearly departed. These covert night-time operations continued to haunt the city, making their way to into 'Madame M's delusion more than a hundred years later, when the network was disturbed by excavations of tunnels for the *Métro*. Perhaps Matthews knew about the project. Perhaps he caught sight of the nocturnal activity and witnessed men shovelling load after load from their carts, and was affected by it, like 'Madame M' when she saw the packing up of First World War uniforms by the side of the road. The catacombs landed a leading role in 'Madame M's delusion as a prison for disappeared soldiers from the First World War and containers of numerous 'lost' children. Whatever the route, the catacombs had also found their way into the image system for the Air Loom, the interconnecting channels corresponding to Matthews's idea of a concealed network of spies. The process of moving the dead was paused during the Revolution, as were Mesmer's demonstrations, but on an early sojourn in Paris Matthews could have stumbled across the spectacle in the dark. Looking back at events from his cell in Bedlam a scene like that becomes a bad omen.

So how does his doctor John Haslam interpret Matthews's allegations? Haslam has known his patient for over a decade by the time he sits down to write a book about him in 1810. He does not dwell on the reasons for the delusion itself, psychological or otherwise. His primary concern is with the question of whether Matthews should be considered legally sane or insane. There is a solid reason for this. Matthews has just been declared of sound mind by a judge after years of legal argument instigated by his family. After he was committed, the family continued to protest on his behalf. They brought a writ of habeas corpus in 1809, and, after long interviews with Matthews, a panel of doctors and lawyers could find 'no evidence of derangement'. With *Illustrations of Madness*, Haslam is attempting more than just the retelling of a page-turning conspiracy story. The work represents a defence of his professional life, and of the overarching clinical approach taken to mental disturbance at Bedlam during his tenure. Haslam prefaces the book with an account of the recent court case which has put his judgement in the spotlight along with Matthew's mental state. Haslam is diplomatic, praising the qualifications of the eminent panel, even though they contradicted his own panel of witnesses. He is also clearly determined that the book should be proof, *in Matthews's own words*, that he is completely out of his mind. He has been mad all along, Haslam insists. He does what he promised Matthews and publishes the exposé of cloak-and-dagger corruption but it's with the hint of a sneer on his lips. Between the lines, he signals that staff at Bedlam were correct in their diagnosis and should be exonerated from any charges of malpractice. Bitterness leaks out between every line and there are a few clear accusations: Matthews's claim that external forces are controlling him is a ploy that would not stand up as a defence in a court of law. It constitutes a failure to take responsibility for his actions and to go along with it would set a dangerous precedent. 'There are already too many maniacs allowed to enjoy a dangerous liberty,' he argues pompously, 'and the Governors of Bethlem Hospital, confiding in the skill and integrity of their medical officers, were not disposed to liberate a mischievous lunatic to disturb the good order and peace of society.'

In *Illustrations of Madness* Haslam's reputation-saving agenda gets in the way of his patient, standing between Matthews and anyone wanting to hear him out. Nevertheless, the relationship between Haslam and Matthews is long and more finely balanced than it first appears. The duel reflects a time when the traditional power dynamic between doctor and patient was unstable. The question of who was mad, who was not and who had the power to decide had always been a hot topic, but it was now the subject of particularly feverish debate. Even the mental capacity of King George III was in question. The issue of how the mentally unbalanced should be treated was also nowhere near decided.

An awareness of delusion as psychiatric illness in its own right began in earnest in France at the beginning of the nineteenth century in the years after the Revolution. Before 1800 or thereabouts, to be delusional was to be inhabited by demons, and the treatment was not medical. It came courtesy of the clergy with the rituals of exorcism. The seventeenth-century philosopher Thomas Hobbes encapsulates the prevailing opinion as essentially binary: 'The opinions of the worlds, both in Ancient and later ages, concerning the cause of madness have been two. Some, deriving from the Passions; some from Daemons, or Spirits, either good or bad, which they thought might enter into man.'[8]

After the Revolution, secular thinking around delusions as a medical disorder began to crystallise. Paris was a natural epicentre for this disruptive new approach. The Revolution had led to the abolition of arbitrary imprisonment – standard practice under the previous regime – and this demanded new ideas on how to treat the mentally ill. They could no longer be locked up on the strength of a cursory letter committing them, and the key thrown away. Something had to be 'done' with them. A cluster of major asylums sprung up at Bicêtre, Charenton and Salpêtrière, which was reserved for women, Maison Blanche and, after 1867, Sainte-Anne. The patients were carefully studied for the first time with a view to recording the symptoms and searching for therapeutic solutions.

The general shake-up of the Revolution freed doctors to work on understanding delusions in a new way: as a psychiatric illness. The pioneers were physicians such as Philippe Pinel, often referred to as the first modern psychiatrist, and his student Jean-Étienne-Dominique Esquirol. They laid down the basic language of psychopathology during these turbulent years.

As a passportless James Tilly Matthews sat in his cell at the Hôtel de Biéville at the height of the Terror, desperately plotting his next move, Pinel was getting his feet under the desk at Bicêtre asylum. Most medical histories include the 1876 painting by Robert Fleury of Pinel heroically cutting the chains of the mad at La Salpêtrière. Although his reputation as the single-handed liberator has been taken down a peg or two in recent decades,

Pinel's leadership in those early days of psychiatry did see innovation in terms of new treatments for the mind. Psychoanalysis did not yet exist, and nor did penicillin, which would have helped many people with delusions as they were symptoms of syphilis which was prevalent in the population at the time, or porphyria, the liver disorder that may have afflicted George III. Nevertheless, this desire to listen to people experiencing delusions marked an important change. Doctors were struggling to understand and patients struggling to be understood.

After Pinel, Jean-Étienne Esquirol continued the work, putting together a long list of delusional disorders, formally categorising them, for the first time, to create a working taxonomy, a map for treatment. 'Delusion of grandeur', 'delusion of doubles', 'delusion of persecution', 'erotomania', 'somantic' delusions involving transformations of the body, to name just a few of the most notorious. French physicians coined one iconic title after another. 'The patients', Esquirol notes in a textbook introducing the idea of 'monomania' – a mode of obsessional thinking related to delusions – 'seize upon a false principle, which they pursue without deviating from logical reasonings, and from which they deduce legitimate consequences, which modify their affections, and the acts of their will. Aside from this partial delirium, they think, reason and act like other men.'[9]

The legacy of this careful attention is the remarkable body of case studies left to us in the archives of the major asylums in Paris.[10] Pinel referred to them as 'anecdotes' and 'sketches' but he knew, as did Esquirol, that these intimate accounts were a treasure trove of primary sources. There were plenty of gaps in the notes, but each case formed a part of an ambitious project. This was the advent of modern medicine, when physicians wrote things up so they could compare notes and improve treatments. An array of medical journals were established around this time to the same end. The spike in reports of delusions at the turn of that century is explained in part by the spike in medical descriptions of all kinds.

In London, by contrast, Bedlam was a worn-out institution in terminal decline, although there was more enlightened thinking emerging elsewhere in Britain. One establishment even threatened to overtake French reform in scope, and in terms of the freedoms that patients were allowed. The Quaker family the Tukes had made money, neatly enough in the tea business, and put their minds and resources to reform in psychiatric hospitals. The patriarch William Tuke, his son Henry and grandson Samuel, established the Retreat asylum in York, an institution that borrowed the phrase 'moral treatment' from Pinel's 'traitement moral' in France. The Retreat adopted a kinder, more thoughtful approach to the custody and care of the mentally ill after the example set in Paris, incorporating into the design extensive open gardens in which the patients could wander and 'lose themselves'. Gardens would become prize features of more and more institutions over the coming decades, such as the one at Vanves, a carefully contrived pastoral idyll outside Paris that 'Madame X' looked out across while explaining to her doctor that she was already dead. Samuel Tuke wrote up a popular account of the York Retreat, sending the vision out into the wider world, and when Bedlam was eventually rebuilt in Lambeth, the superintendent visited the Retreat for inspiration and admired its atmosphere.

Matthews remained in Bedlam for almost the rest of his life, tending his own little patch of soil there, but the family as a whole never lost faith in him over the decades he remined in the institution. There was a nephew by marriage who kept up the pressure along with Elizabeth Sarah. The tireless years of litigation came at great expense, especially against the likes of Haslam who had a strong incentive to fight them at every stage in order to save his reputation.

Jamaican parish records show Elizabeth Sarah emigrated there with their child at some point, after years petitioning for her husband's release. She may simply have wanted to escape the reputational damage of being married to an inmate at Bedlam. More likely she was desperate to relieve the financial pressure on her. There were servant positions to be had in the busy colonial estates. Having maintained contact for so long it's a painful fact she was never allowed back inside to see him. The last image we have of her is a long way from Camberwell and even further from the fine church in Bloomsbury where she married James.

The family managed to bring one more case to court to argue that Matthews was sane and should be allowed to leave Bedlam. Tenacity was clearly a family trait, and it finally paid off. Matthews was declared sane and released on 27 June 1813. He moved into a private establishment in Bethnal Green into the care of a Mr Fox. Fox apparently agreed that he was in his right mind and he was much freer to come and go as he pleased. He was even allowed to carry on helping with the gardening as he had done at Bedlam.

After a decades' long duel with Haslam, Matthews delivered a devastating parting shot. His testimony about abuses at Bedlam to a Select Committee for a report on 'Madhouses' convinced the panel, and helped to see Haslam dismissed after its publication in 1816. His words contributed to reform at the institution.

And what of Matthew's delusional beliefs? Did they lose any of their grip following the recognition and validation given to him by the court case and its verdict? Or did he conclude they believed his allegations, might even foil the plot thanks to him? Matthews died in Mr Fox's house less than two years after he arrived, on 10 January 1815, aged forty-eight. The clerks at Bedlam cared enough about their former inmate to record the event in the log.

The hall of fame of paranoid delusion has collected many other remarkable characters since James Tilly Matthews. Another landmark case was that of Daniel Paul Schreber, a judge in *fin de siècle* Germany. His case was immortalised by Freud in 1911 who gave a commentary on Schreber's 1903 *Memoirs of My Nervous Illness*. The action focused on Schreber's doctor, Professor Paul Flechsig. Schreber came to believe that his psychiatrist was controlling his thoughts and actions, using a 'nerve language', a complex communication system which linked souls to God. Others were able to make contact with his nerves through 'divine rays'. He also believed that Flechsig was trying to transform him into a woman. Schreber had intended his memoir to interrogate the question of legal ethics around keeping someone inside an asylum against their will. Freud glossed the delusion differently: Schreber's repressed impulses and ambivalent feelings about his brother and father were leading him to project motives onto his doctor, Flechsig, even onto God.

There are numerous, quieter cases of paranoid delusion to be found in the archives. Ellen Hamilton, for example, a forty-one-year-old woman from St John's Wood, London, who was admitted to Bedlam on 10 January 1892, nearly eighty years after Matthews was released into the private care of Mr Fox. In her admission case notes, Hamilton is greatly distressed in her belief that people are 'telephoning into her ears' and that 'people had extorted money from her by means of the telephone, that her house had been blown up… Susan Glynn, sister of above… told me that her sister sometimes walked about all night, that she accused people of stealing her property, that she heard voices through imaginary telephones making unpleasant accusations.'[11]

Paranoia and conspiracy are the defining delusions of the last hundred years. Nanotechnology, social media and reality television have replaced pneumatic chemistry and mesmerism as conductors of threat at this point in the twenty-first century.

After 1945, discoveries of spying apparatus regularly made the news and in the US during the Cold War countless people became convinced the security services were downloading their thoughts through micro-transmitters: they were reading minds. 'The Great Seal Bug', a piece of spy kit otherwise known as 'The Thing', was the most infamous. A device for covert listening, it was discovered in 1952 and eventually made public in May 1960 when the story was splashed across the international press. This was a carved wooden wall plaque, a copy of Great Seal of the United States, given as a gift from the Soviet Union to the US ambassador in Moscow, W. Averell Harriman, on 4 August 1945. The bug was secreted inside and it remained on the wall in the Moscow study for seven years. Eventually, a British radio operative intercepted a conversation and realised the Soviets were beaming radio waves at US diplomatic buildings. The device functioned in a completely novel way. It was active only when a strong electromagnetic signal of the correct frequency was sent to it from outside to wake it up. It had no power supply so had a potentially unlimited life, and it was very small and easy to hide. *Get Smart*, a popular television show in the US in the mid-1960s, featured a counter-intelligence office, Maxwell Smart. Smart used a 'shoe phone', a cutting-edge invention with a secret receiver in the heel and a transmitter in the sole that looks a lot like an early mobile. The press shot of Smart holding the shoe to his ear, furrow-browed, eyebrow raised, is a comic pose from light entertainment, but it also signals a growing unease around who might be listening, and how.

The most powerful forces are still the unseen ones, wanting our money or our vote, anonymous behind the screen, disrupting our ability to authenticate facts and identities. With the development of nanotechnology, the idea of a chip in your tooth downloading your thoughts is not as implausible as it was three hundred years ago, or even five years ago. It's increasingly tricky to separate out the false beliefs from the advanced tech. Wireless devices in our home listen to us, collect data about us and adjust their algorithms accordingly. The machines really are listening.

A delusion of conspiracy organises the enemy, gives a shape to ambiguity, a face to anonymous players and fears. It gives a person a job to do: a bad guy to fight, and, by its own logic, casts the person experiencing it as the good guy, legitimising any subsequent actions.[12] Any evidence that a belief isn't true is further proof of conspiracy. The belief won't answer to reason because it's working very successfully on its own terms, and there are compelling reasons for a person to keep it up. In his 1974 paper, Harvard psychologist Brendan Maher said, 'the delusional belief is not being held in the face of evidence normally sufficient to destroy it, but is being held because of evidence powerful enough to support it'. Many delusions represent a certain simplification of the world: a world which is easier to navigate, somewhere a person can feel safe, where they play an important role. Conspiracy theories often represent a dangerous oversimplification. It's not an easy sell to coax a person back from a gloriously neat world into a more nuanced and confusing one.

There are many gaps in the records into which we lose aspects of Matthews the man. We can see him more clearly in relation to others. The way his delusion functions, at its simplest level, connects him to people who have also struggled to contain incompatible demands, shocks and falls and forged an alternative reality to live in. He maps the world for himself, describing the landmarks and the personnel. He has agency, and his role is of critical importance.

At the centre of Matthews's delusion is something more original: a strange and intimidating machine. The Air Loom lives up to its name, looming down at us as a premonition of the mechanised world of the nineteenth century, and the technology of the twentieth, which will see people, even God, demoted. Matthews performs like a medium at a seance, his visions anticipating the darker and deeper conspiracy theories that will proliferate over the following centuries. A more personal story comes through, too: here is a self-made idealist who experiences humiliation, rejection, feelings of powerlessness. His is a psyche under extreme stress, pulled in too many directions; a peacemaker who cannot make peace, who finds himself unwelcome everywhere he turns. He brings an artist's imagination to the problem and the Air Loom is the artwork.

Robert Burton portrait in oil by Gilbert Jackson, 1635, in
the collection of Brasenose College, University of Oxford.

CHAPTER III

The Melancholic Delusions
of Robert Burton

25 January 1640, Christ Church College, Oxford. A sixty-three-year-old man in a senior scholar's gown, mortar board and ruff sits at his desk in his well-positioned rooms off the main quad, his topiary beard forming a promontory, writing something important which requires his full attention. He is taking his time, concentrating on the task with doleful brown eyes over an aquiline nose and a faint smile. If you could look over his shoulder you would see that these are the words for an epitaph. It's his own tribute, and it's to be displayed in the north aisle of the cathedral at Christ Church after his death, beneath his likeness, a bust, which will be brought to life in naturalistic paint colours.

Now he carefully begins to transcribe something: a horoscope. He calculated it for himself some years before, capturing the stars' alignment at the time of his birth. He leaves instructions that these documents should go to his elder brother, William, who will see to their commission when the time comes.

He removes his mortar board and gown, and the starched white ruff, then takes the embroidered runner from his desk and fashions it in such a way that it creates a slip knot. This serves as a noose which he places around his neck. And then he hangs himself, there in the very room in which he has spent a working

lifetime revising his magnum opus – an encyclopedia of melancholia and associated delusions.

Anyone who stumbled on the scene, and could read the
geometry and symbols in a horoscope, would see that according to the diagrammatic representation on the desk Burton's
climacteric year (sixty-three) would be his last. It would 'prove
fatal', as the saying went. Once they learned that the dead man
was sixty-three years old, they would have to conclude that he
had died at the age predicted. The forecast had been accurate,
or at least Burton had believed it so strongly that he had taken
it upon himself to prove it correct. Everyone said he had been
in good health up until that day. So the rumours went, anyway,
disseminated in whispers by the students.

By the time of his death, the obsessional Oxford don Robert
Burton had assembled what is still to this day the most important
compendium of mental disturbance and accompanying delusions:
The Anatomy of Melancholy, first published in 1621 in several volumes.[1] In his book Burton was open about the fact that he suffered
with bouts of melancholy, but did a learned man like him really
believe that a horoscope had prefigured his death? Is it possible
that he even brought the prediction about on the appointed day?
He's certainly the subject's great collector, but do we have a case
of delusion in Burton himself?

Burton is writing for an early modern Christian audience,
but he draws extensively on classical sources for his case studies,
as well as on examples from his own day. It is because of his
magpie-like collecting habits that we still have such early delusion cases to refer to at all. Many of these stories had already
been recycled over hundreds of years when Burton fell upon

them (he didn't travel far in terms of distance – these were voyages around the bookshelves of England: he 'turned over such physicians as our libraries would afford, or my private friends impart, and have taken this pains'.[2] The most intriguing stories from any era which caught his eye made it into this book. The title page announces an exploration of melancholy 'with all the kinds, causes, symptoms, prognostics, and several cures of it… philosophically, medicinally, historically opened and cut up'. This remains the organising work on a sprawling topic. He writes of the myriad ways in which the 'chaos of melancholy' symptoms express themselves. 'They will act, conceive all extremes, contrarieties and contradictions, and that in infinite varieties… Scarce two of two thousand that concur in the same symptoms. The Tower of Babel never yielded such a confusion of tongues, as this chaos of melancholy doth variety of symptoms.' But he chases them, a maelstrom of reference, determined to guide them into order. The *Anatomy* became a publishing sensation within his lifetime, and built a readership well beyond it. The fancy, and pricey, folio made eight editions over the course of the seventeenth century alone. Laurence Sterne borrowed entire passages for his novel of 1759, *Tristram Shandy*. In 1650 the astrologer and influential almanac producer John Gadbury was reading it in bed when a reference to horoscopes jumped out at him and decided for him what he would do for a living.

The rumours as to the mode and meaning of Burton's death persisted. Speculation was rife in Oxford for decades, and repeated through several channels. Anthony à Wood, a chronicler of Oxford life writing fifty years after the event, reported that the students still gossiped about how Burton had died. They said it was 'at, or very near the time, which he had some years before foretold from the calculation of his own nativity'.[3]

According to William Tegg's 1863 edition of the *Anatomy*, which quotes Wood extensively to borrow some of his colour, it went like this: 'In his chamber in Christ Church College, he departed this life, at or very near the time which he had some years before foretold, from the calculation of his own nativity, and which, says Wood, "being exact, several of the students did not forbear to whisper among themselves, that rather than there should be a mistake in the calculation, he sent up his soul to heaven through a slip about his neck."'[4]

The biographer John Aubrey, picking up the tale later in the seventeenth century, offers the following version of events via Robert Hooke, the very same Enlightenment polymath who designed the Bethlem hospital at Moorfields. 'Mr Robert Hooke of Gresham College told me that he lay in the chamber in Christ Church that was Mr Burton's, of whom 'tis whispered, *non obstante* all his astrology and his book on Melancholie, he ended his days in that chamber by hanging himself.'[5] The insinuations in the italics here are whispered behind discreet hands.

Robert Burton adopted a *nom de plume* for the *Anatomy* and stamped it across the title page. This name was 'Democritus Junior'. The Greek philosopher he chose to hide behind was famous for finding the follies of the world so absurd that all he could do was laugh. How serious is Burton being?

The plaque beneath the memorial bust of Burton on the wall of Christ Church, Oxford, bears the epitaph he requested, inscribed by his brother as per his instructions. It's an enigmatic signing off:

Memorial bust of Robert Burton, Christ
Church Cathedral, Oxford.

*Paucis notus, paucioribus ignotus, hic jacet Democritus Junior, cui vitam
dedit et mortem Melancholia*

This translates as: 'Known to few, unknown to fewer, here lies
Democritus Junior, to whom Melancholy gave life and death.'

The horoscope has been painted onto a gilded boss to the right
of the bust. It is a geometric image of intersecting diamonds, the
lines radiating out like a star, with the date of 8 February 1576
at the centre. Around it numerals and astral positionings hang
in delicate hand-scribed calligraphy. A decorative portrait of his
fate. The public position of this horoscope ensures that he will be
associated with it for ever. Burton's elder brother saw to all the
arrangements but the orders were all Burton's, Anthony à Wood
confirms. This was a man hooked on astrology.[6]

The opening to the epitaph is playful. He refers to himself as 'known to few, unknown to fewer', acknowledging with a coy smile that he may be awfully famous, but he is also a man of such complexity and mystery it's impossible to understand his hidden depths. Another joke? There's more. The epitaph goes on to name the other guiding force in his life and death, melancholy. He has been its child and its slave, always under its control. Melancholy and astrology are his twin masters. This memorial represents his life pared back to the key structural elements. It publicly ties his mental state to horoscopes. It's a sombre admission.

It seems he did believe, but a renowned scholar like Burton telling the world he is under the influence of the stars is risking his academic reputation. Where did this conviction come from? At first glance, the belief declared on the wall at Christ Church Cathedral seems an unlikely one for a man in his position. What's it offering him?

Burton offers an intimate portrait of his daily routines in his preface to the *Anatomy*. He is in his mid-forties, living a 'silent, sedentary, solitary private life...penned upmost part in my study'.[7]

> ... I live still a collegiate student, as Democritus in his garden, and lead a monastic life, sufficient entertainment to myself, sequestered from those tumults and troubles of the world... I hear and see what is done abroad, how others run, ride, turmoil, and macerate themselves in court and country, far from those wrangling lawsuits, I laugh to myself at the vanities of the court, the intrigues of public life, I laugh at all...those ordinary rumours of war, plagues...massacres, meteors, comets... heresies... Today we hear of new lords and officers created,

to-morrow of some great men deposed...amidst the gallantry and misery of the world... I rub on in complete privacy; as I have still lived, so I now continue, left to a solitary life and mine own domestic discontents: saving that sometimes, not to conceal anything, as Diogenes went into the city and Democritus to the haven to see fashions, I did for my recreation now and then walk abroad, and could not choose but make some little observation, less by way of shrewd remark than of simple statement of fact, not as they did, to scoff or laugh at all, but with mixed passion.[8]

With a wink he admits to a little 'recreation' time, but only to emphasise that he doesn't enjoy it too much. We are invited to picture him, in self-imposed isolation, living an ascetic existence, hunched over his desk, researching and revising. He is trying to define the societal scourge of 'melancholia' and its attendant delusions for the public: ruminating on the causes, setting down the symptoms and dedicating the whole of the second part to possible cures.

Anthony à Wood paints a rather different picture. His Robert Burton is a rather jolly sort. Wood apparently 'heard some of the ancients of Ch.Ch. often say that his company was very merry... and juvenile, and no man in his time did surpass him for his ready and dextrous interlarding his common discourses among them with discourses from the poets or sentences from classical authors. Which being then all the fashion in the university, made his company more acceptable.'[9] He was the life and soul of the Common Room classics party.

The English diarist and antiquarian Thomas Hearne corroborates this version of Burton's character (although Hearne wasn't born until after Burton's death, and does not quote his source. He may have been relying on Anthony à Wood). 'Mr Burton was one of the most facetious [jolly] and pleasant companions of that age, but his Conversation was very innocent. It was the way then to mix a great deal of Latin in discoursing, at which he was wonderfully ready (in the manner his book is wrote) which is now looked upon as pedantry.'[10] His party trick of slipping between Latin and English seems to have gone down a storm. Burton's posture as a recluse appears even more misleading in light of the fact that he was at one point a college librarian as well as a vicar – he even joined the fray for a period as clerk of the busy Oxford market.

College rules did not allow him to marry, but a Latin poem prefacing the *Anatomy* divulges the author's fondness for any dainty damsels who might read his book and whom he loves 'dear as life…would he were here to gaze on they sweet look'. While he 'never travelled but in map or card, in which mine unconfined thoughts have freely expatiated'. He stayed put, but went anywhere and everywhere in his imagination.

There are key moments in the sprawling text of the *Anatomy* where Burton addresses the matter of his belief in astrology directly. In one scathing aside he warns his reader to be sceptical and beware of charlatans: 'what is astrology but vain elections, predictions? All magic, but a troublesome error, a pernicious foppery?'[11]

But at another point, he is ambivalent about the merits, both in terms of what were termed its 'naturall effects' in areas such as the weather and disease; and its 'judicial' or predictive powers in the form of horoscopes. He claims diplomatic neutrality on the matter. 'I will not here stand to discuss *obiter*, whether stars be causes, or signs; or to apologize for judicial astrology.' He doesn't want to be dragged into the astrology-as-science-or-quackery fight, let alone to arbitrate, but he lists the defenders as well as the attackers. Certain figures, Sextus Empiricus, Picus Mirandula, Chambers, for example, present the spheres on an astrological chart as no more significant than a street hoarding 'he will attribute no virtue at all to the heavens, or to sun, or moon, more than he doth to their signs, at an innkeeper's post, or tradesman's shop'. But some prominent physicians successfully use the stars to predict a person's susceptibility to certain diseases; there's Bellantius, Pirovanus, Marascallerus. And, of course, then there is the respectable public figure of Christopher Haydon to consider. Haydon is a Member of Parliament, and a defender of astrology as a science that is not only valid but entirely compatible with Christianity. Just as he seems to be about to side with the astrology-as-science lot, he steps back again. Regarding the stars, 'If thou shalt ask me what I think, I must answer, for I too am conversant with these learned errors, they do incline, but not compel; no necessity at all…and so gently incline, that a wise man may resist them…they rule us, but God rules them.'[12] The stars 'do incline, but not compel,' he concludes, tactfully. The influence is so subtle that missing the evidence doesn't make you stupid. Burton finds a vantage point with a strategic view. It's bifocal: a belief in the predictive power of astrological forces running in parallel with a strong Christian faith. The stars rule us, but God rules them. God is still at the top. Burton wanders away from the topic again into the shadows.

Elsewhere in the *Anatomy* he comes out of hiding for a very candid outburst on the subject. This is a warning about the dangers of learning about your future. You can, he warns, make something come true just by fearing it:

They are so much affected that, with the very strength of imagination, fear, and the devil's craft, 'they pull those misfortunes
they suspect upon their own heads, and that which they fear
shall come upon them,' as Solomon foretelleth… 'The thing
that I feared,' saith Job, 'is fallen upon me.' … the foreknowledge
of what shall come to pass crucifies many men… There was a
fountain in Greece, near Ceres' temple in Achaia, where the event
of such diseases was to be known… Amongst those Cyanean
rocks at the springs of Lycia, was the oracle of Thryxeus Apollo,
'where all fortunes were foretold, sickness, health, or what they
would besides': so common people have been always deluded
with future events…if he foretell sickness such a day, that very
time they will be sick…and many times die as it is foretold.[13]

Burton grabs his audience by the shirt collar. This is not a rhe-
torical set piece, it is the kind of zealous advice forged from lived
experience. He is positively evangelical about the dangers of
becoming 'deluded with future events'. He cites Job, the Bible's
most famous victim of a celestial wager about the strength of a
man's faith, as well as those poor souls destroyed by what they
learned at Apollo's Oracle. In a sense he implicates all futurology,
both before and after his time, from the Druids reading entrails
to twentieth-century tea-cup twisters reading the leaves, to palm
and tarot card readers. He is well versed in how the imagination
can get the better of a person when they hear what's supposedly
in store for them, and knows that the consequences can be deadly.

But Anthony à Wood repeats the rumours that Burton didn't
heed his own advice. He was 'by many accounted a severe student,
a devourer of authors, a melancholy and humorous person; so by

others who knew him well, a person of great honesty, plain dealing and charity'. But this esteemed and sensible scholar was, on occasion, a 'curious calculator of nativities'. Burton couldn't stay away, it seems. He couldn't resist looking into the future, even though he understood this knowledge had the power to ruin a person.

It's a talking point on the streets. In 1605 or thereabouts, while Burton was working away at Christ Church and struggling to take a definitive line on astrology, another writer was offering his assessment to theatre audiences. Shakespeare's *King Lear* is a savage dismantling of the credulous enthusiasts of astrology. Edmund, the malcontent illegitimate son of Gloucester, mounts a critique of believers like his father. Gloucester has been wringing his hands about how eclipses portend bad things. His son rails: 'This is the excellent foppery of the world that when we are sick in fortune – often the surfeit of our own behavior – we make guilty of our disasters the sun, the moon, and the stars, as if we were villains by necessity, fools by heavenly compulsion, knaves, thieves and treachers by spherical predominance, drunkards, liars and adulterers by an enforced obedience of planetary influence.' Edmund charges on, Shakespeare lacing his words with dark sarcasm: 'My father compounded with my mother under the dragon's tail and my nativity was under Ursa Major, so that it follows I am rough and lecherous. Fut, I should have been that I am, had the maidenliest star in the firmament twinkled on my bastardizing.'

In the wider culture, the astral code is starting to seem quaint. The power of the stars is being questioned in public. Believers are lampooned on the London stage as gullible. It would be more tempting than ever for a person who kept faith with the old celestial principles to keep the fact a secret.

There are a few moments in the *Anatomy* where Burton explicitly attributes his own melancholy to the arrangement of these stars, and several volumes belonging to Burton in the Bodleian Library in Oxford show that Burton was never really on the fence. One of these volumes is marked with his autograph, and the date 1603, when he was twenty-six. The margins of the books are filled with his annotations. These annotations all relate to astrological practice. The notes in his copy of Stadius's *Ephemerides*, a handbook for astrologers for the dates 1554–1606, cover the period he was a student at Oxford and in his twenties. There's also a treasured edition of Ptolemy. These are detailed scribblings about precisely of the sort of 'judiciall' astrology he warns against in the *Anatomy*.[14] A notebook includes tables as well as related jottings on astrological topics. There's even a horoscope for Queen Elizabeth I in there. This notebook is also where he drew up his own horoscope. He notes down his birth, at 8.44 a.m. on 8 February 1576. This matches what's written above his epitaph in Christ Church Cathedral. He's also recorded the time and date of his 'conceptio', or conception. This is down as nine o'clock in the evening of 25 May 1575. If his parents shared precise details like this with their son, the suggestion is that they, too, had an interest in horoscopes.

Astrology was still popular, both for its 'naturall effects', weather and disease forecasting, and its 'judiciall' horoscopes, and an astrologer might be called in not only to forecast the life of an infant, but to learn the whereabouts of a missing person or property. The practice of future-telling by the arrangement of the stars was by definition at odds with Catholic teachings in that it removed the onus on an individual to lead a pious life in the hope of salvation. Burton was in good company, though, in accommodating devout Christian faith with an interest in astrology. Prophetic almanacs were published annually and sold in large numbers and almost everyone would have been going to church.

By the sixteenth century the Oxford colleges were flourishing in terms of both population and prosperity, having undergone a revival during the reign of Elizabeth I. James I visited Oxford while Burton was ensconced at Christ Church, in 1603 and 1605, and he and his queen were presented with cups of money.[15] Plague menaced the city on and off, a major outbreak occurring in 1603, when the churches were closed and grass grew in the market place. In 1593, precautionary measures sanctioned by the mayor included the removal of rubbish and pigs from the streets, restrictions on lodging strangers, especially from London, and a ban on plays. The mass gatherings at the university were blamed for the outbreak of plague that year.

The real tension in Burton's Oxford, though, was between the old and new church, and as vicar of St Thomas's he was in the thick of it. Puritan elements existed at the university from the 1580s, and by the early seventeenth century the practices in some of Oxford's churches were changing: taking sacrament sitting down, for example, instead of kneeling. There was opposition to Whitsun and Mayday sports and celebrations. There was also a revival of the old church, with crucifixes on show in some of the city's churches. There was even a statue on the porch at St Mary's, which was cited as proof of Archbishop Laud's popery at his trial. Burton seems to have been part of this 'Laudian' movement, along with others who set themselves against the Puritans and rejected the idea of predestination, finding common ground with the Roman Catholics. Burton was an Anglican, but kept up the Catholic practice of using unleavened wafers for Communion. Burton's own church, St Thomas's, was built on land seized from Osney Abbey at the dissolution of the monasteries earlier in the century and given to Christ Church. Cranmer, Ridley and Latimer had been burned in public in Oxford in 1555/6 in the throes of the Reformation. When Elizabeth took the throne after her Catholic sister Mary, most parishes in the city conformed to the settlement and changed their rituals and furnishings once again, but some hedged their bets and kept the vestments and candles just in case.

By the time Robert Burton began collecting material for *The Anatomy of Melancholy*, James I was on the throne and religious authorities had ordained a direct link between delusions and demonic possession. Delusions as they appear in the *Anatomy* are caught in a web of philosophical and religious treatises, medical theory and poetry; superstitious, humoral, Christian and psychological perspectives. That's where most people were, too.

Burton's Christ Church was a major centre for mathematical study. Few scholars there would have called astrology a delusion. It was too closely allied to the cutting-edge science of *astronomy* and the rate and scale of the discoveries in that field required a degree of magical thinking anyway, just as the new forces emerging in James Tilly Matthews's day had.

The revolution in astronomy was gathering pace as Burton was researching, publishing and revising. Western Europe was still reeling from Copernicus's theory of 1543 that the earth revolved around the sun. In 1572, just before Burton was born, a brilliant 'supernova' was spotted by Danish astrologer Tycho

Brahe. Brahe proved that the star was moving beyond the earth's atmosphere and, therefore, that the heavens could change positions. In 1608, when Burton was thirty-two, the Dutch eyeglass maker Hans Lippershey produced a refracting telescope. The invention spread rapidly across Europe, with scientists making their own versions, and then German astronomer Johannes Kepler's laws of planetary motion showed that orbits were elliptical. In 1610 Galileo Galilei published *Sidereus Nuncius*, describing his observations with a retractable telescope that offered even greater magnification. Galileo backed 'heliocentrism', enraging the church by contradicting the belief that all heavenly bodies revolved around the earth.

The planets must suddenly have appeared even more powerful forces than anyone had previously dared to imagine. Galileo pointed out the craters on the moon, stars that could not be seen with the naked eye and even moons circling Jupiter. As Burton developed his understanding of melancholy, new insights about the sky above were swirling around him.

The reference points of astrology had not changed at all for centuries, and offered continuity and reassurance. Astrologers only referred to the planets that were known about at that time: Mercury, Venus, Mars, Jupiter and Saturn. No new planets were discovered until 1781 when Sir William Herschel found Uranus. Later in the century the English astronomer and mathematician Isaac Newton demonstrated the physical laws which acted on the planets to define their orbits and dismissed the validity of astrology out of hand. He created a reflective telescope with a mirror instead of a bulky glass lens, making it smaller and more accurate so that more people could have a closer look at the night sky and see for themselves. He was, though, a deeply religious man and saw theology and maths as one project to discover a single system of the world. His secret hobby was alchemy.

When Robert Burton was a boy, in the late sixteenth century, astrology was still a respectable science and an openly popular hobby.

Burton's birth is not in the official records, but all circumstantial evidence points to the morning of 8 February 1576. He was the fourth of nine children, and the younger of two sons, born into gentry from Lindley in rural Leicestershire. The family lived in the manor house, Lindley Hall. Robert's brother William was an antiquarian who wrote a historical and topographical survey of the county entitled *Descriptions of Leicestershire*, which he began in 1597 and published just after the *Anatomy* in 1622. On the very first page of William's book he says that he hopes to shine light on the county, by removing 'an Eclipse from the Sun, without Art or Astronomical calculation'.

Perhaps the family read almanacs, as many families did, or even consulted an astrologer in person for a fee in exchange for advice or answers to questions of the future.

The manor house in which the Burton boys were born is described as a 'curious mansion with turrets entirely surrounded by a moat, excepting a way or road to the house, and surrounded by an avenue of trees'.[16] The turreted seclusion is a rehearsal for his later cloistered life at Oxford. The house was, by the sound of it, completely cut off from the outside world by a watercourse, with acres of inherited land wrapped around. Any prospective visitors had a real challenge getting a horse and carriage to the house; the manor was knocked down in the eighteenth century and the surroundings re-landscaped to allow access and to correct the blight. The Burtons had everything they needed, their own little universe, at Lindley: orchards, stables, even a chapel. Both boys would have had a lot of time to consider the twin sciences of astronomy and astrology, and looking out of their windows at the night sky it would have been hard to separate one discipline from the other, or the science from the magic.

History was very much alive around Lindley. In his homage to Leicestershire, William proudly notes the excellent soil and pastureland, as well as a blessed lack of snakes, and he takes a moment to boast direct Burton family lineage to a nobleman who accompanied Richard I to the Holy Land. The Lindley manor sat near Bosworth Field, where Richard III met his death, and the graves of many who died in that battle are in the local churchyard. Given that they kept hold of their estate after the Reformation, it seems the family stayed on the right side of religious authorities over the decades of flip-flopping to and from Rome. The association of the family with Arthur Faunt, a Jesuit relative of Burton's mother Dorothy, hints at Catholic sympathies. Dorothy was apparently skilled in 'chirurgery' which involved minor surgery and bone setting, and perhaps the medical interests of the younger son were learned at his mother's knee. Anthony Faunt, another maternal relation, was said to have died from 'a passion of melancholy'.[17]

Burton did finally leave Lindley to go to the local grammar school in Nuneaton and later in Sutton Coldfield, just over the county boundary in Warwickshire, where the sciences of astronomy and astrology would both have been on the curriculum. He followed William up to Brasenose College, Oxford, and then to Christ Church to complete his first degree, where he remained.

And it's at some point after he starts at Oxford that a brooding relationship with melancholy, the other defining influence on his epitaph, takes hold of him. He was enslaved by it, he says. In the *Anatomy* he will be good company, gossipy and humorous, but then slam a home truth in front of his reader:

'All the world is mad… it is melancholy, dotes… 'Tis an inbred malady in every one of us.'[18]

He is himself 'melancholy' while at Oxford. The standard age to graduate at that time is nineteen, but Burton is twenty and still at Oxford and he will not graduate until he's twenty-six. There is a gap in the records between 1593 and 1599 which may be explained by the depressive illness that's stalking him.

Pieter Codde's portrait of a boy, *Young Scholar in His Study, Melancholy*, could be a young Burton staring blankly back at us:

Young Scholar in His Study, Melancholy,
Pieter Codde, c.1633.

'Melancholia' is the thread linking the classical world to the anxieties and preoccupations of Burton's own day. The concept had already endured over a couple of millennia. Albrecht Dürer distilled its elements in a dramatic engraving featuring a winded personification of Melancholy, staring despondently out into the world, surrounded by astrological instruments. Next to her sits an hourglass, reminding us that time is running out. A comet and a rainbow appear in the sky above.

Burton's Europe was experiencing a new wave of this old affliction. It could be debilitating but was also, at this point, achingly fashionable. Melancholia was something of an epidemic in Europe in the sixteenth and seventeenth centuries; a distracted gaze off into the middle distance, as you recline by a babbling stream, was a sign of the intellectual refinement which might get you a diagnosis of 'Scholar's Melancholy' (it was also, of course, a sign of money. The rich had more time to be idle, and could afford a physician to record a case in the first place.). Nicholas Hilliard's moody portrait of Henry Percy, 9th Earl of Northumberland of c.1594–5 showing him lying on his side, propping his head in his hand with a book at his side, is central casting for a melancholy scholar.

The terms used in antiquity to describe mental disturbance do not transpose neatly to a modern world which frames things in psychological or neurological terms. But the idea of 'melancholia' in classical medicine encompassed what we would think of now as depression, as well as delusions. The classical definition of 'melancholia' was something like 'sadness', but it referred to a severe mental disorder rather than the wistful navel-gazing the word can suggest to the modern ear.

Henry Percy, 9th Earl of Northumberland,
Nicholas Hilliard, miniature c.1595.

Hippocrates diagnosed the 'melancholic' as a person with an excess of one of the four basic 'humors' – the elements within the human body that determined a person's temperament and health. In this case, a melancholic patient had too much black bile (μέλας melas: χολή kholé dark/black) and this led to a mental state, above all, characterised by fear and sadness. He defined 'melancholia' as a catch-all for mental disturbance, and put what we would think of now as delusions within its remit.

Galen of Pergamon, a physician working in the Roman Empire, developed the thinking around melancholia and argued that, along with depressive symptoms, melancholic patients showed bizarre and fixed ideas. 'There are patients', he writes in *On Affected Parts*, 'who think to have become a sort of snail so that they must escape everyone in order to avoid having their skull crushed.'[19] He also wrote about a man gripped by the belief that Atlas, the Titan god and Bearer of the Heavens, is tiring and might drop the world that he had been holding up.

The most evocative classical medical descriptions of delusion were laid down by a contemporary of Galen, Aretaeus, a Greek physician working in the Roman province of Cappadocia (modern Turkey). In his textbook *On the Causes and Signs of Diseases* he gives examples of melancholic cases where 'The patient may imagine he has taken another form than his own. One believes himself a sparrow; a cock or an earthen vase; another a God, orator or actor, carrying gravely a stalk of straw and imagining himself holding a sceptre of the World; some utter cries of an infant and demand to be carried in arms, or they believe themselves a grain of mustard, and tremble continuously for fear of being eaten by a hen; some refuse to urinate for fear of causing a new deluge.'[20] The severe mental disturbance described here – nightmares of delusionary hallucinations – are a version of melancholia about as far away from the self-indulgent longing evoked by Romantic poets as it's possible to be. These people have underdone extreme bodily transformation, finding themselves with exaggerated powers and a responsibility to save the world, or rendered powerless as a baby, small animals, even animal fodder.

Again and again melancholics are discussed in relation to their delusions largely because melancholia was believed to damage the imaginative faculty of the brain. It's also because they are such bizarre and enchanting stories.

We can allow ourselves a little Burtonian digression to note here that Arateaus was the first to describe those who 'fear that people wish to give them poison and who develop hatred for mankind, flee into solitude or become surreptitiously addicted to religious practices'.[21] He names paranoia and delusions of persecution and religiosity, demonstrating just how eternal the most common delusion types and themes have proved to be.

Burton championed a new enthusiasm for going back over classical records, and brought cases from antiquity to the sixteenth-century audience. He throws another story into the mix of a man 'that thought he had some of Aristophanes' frogs in his belly, still crying *Brecececex, coax, coax, oop, oop,* and for that cause studied physic seven years, and travelled over most part of Europe to ease himself'.

One of Burton's sources tells of a baker in the Italian city of Ferrara who succumbed to melancholy and became convinced that he was made of butter, so that he dared not go near his own oven or sit in the sun in case he melted. Louis XI of France was convinced that everything around him stank, and not even the sweetest smelling perfumes provided for him by doctors could persuade him otherwise. How could Burton resist stories like these? How could anyone? He attempts to coordinate an unwieldy wealth of material and knowledge and create a working compendium, a companion for fellow sufferers into which future generations could dip for help. He spent ten years or so poring over source material before he put pen to paper, and there is a strong sense that his handwriting can't keep up with the burst dam of accumulated knowledge that continues to spill out.

The *Anatomy* claims to be a clinical enquiry into a distinct mental disturbance, its causes and cures, but it reads something like a stream-of-consciousness with its disorientating breadth of reference and digressive flourishes – a bit of Shakespeare here, a snippet of Chaucer there – and multiple self-contradictions. Burton continues wrestling the chaos into some order and it's harder and harder to tell if the chaos is inside or outside of himself.

In a twist on the convention, Burton addresses his thoughts on melancholy directly to us. 'Thou thyself art the subject of my discourse,' he tells us, turning the mirror on the reader. Burton asks us to do something potentially uncomfortable: look at our own capacity for delusion.

He dedicates the book to his readers, but he perseveres with the project to keep himself busy. Idleness, he explains, is the chief driver of melancholy. 'And these my writings, I hope, shall take like gilded pills, which are so composed as well to tempt the appetite and deceive the palate, as to help and medicinally work upon the whole body; my lines shall not only recreate but rectify the mind.'[22]

The Ashmole Collection at the Bodleian Library in Oxford keeps another notebook significant with regard to Robert Burton and his state of mind. It suggests that as a young man, Burton, an enthusiastic advice-peddler in later life, went to someone else for help. The notebook belongs to one of the biggest stars in the firmament of astrology: Simon Forman. Forman was a renowned physician as well as an astrologer, who practised in and around London in the 1590s and 1600s. The appointments diary for 18 June 1597 records a consultation at 5.40 p.m. Forman received a visit from a 'Robart Burton of 20 yeres'.[23]

Robert Burton from Lindley's date of birth makes him around the right age for him and 'Robart Burton' to be one and the same person. Other evidence is circumstantial but compelling: letters suggest that our Robert Burton visited his brother William, who had been admitted to the Inner Temple, in London around this time.[24]

This 'casebook', one of hundreds belonging to Forman in the Ashmole Collection, is a well-thumbed, working notebook. It's a piece of daily ephemera belonging to a busy astrologer – he gets through them quickly – and it's covered in Forman's spidery hand and difficult to decipher.

But there's his name, next to all sorts of what looks like code related to what transpired when Burton came to see Forman on 18 June 1597. The consulting room was in the Stone House, Billingsgate, on the north side of the Thames in the City of London, the street next to Pudding Lane that will be destroyed by the Great Fire. Elizabeth I is still on the throne.

Forman is more than twice the age of the young man across the desk. He has a full beard and a very prominent brow. As a major celebrity in the field, he's an intimidating proposition. He's also famous for curing himself of the plague which went round London four years earlier. Forman performs two thousand of these consultations a year, so the encounter is bread and butter to him. He is a society fixture, who associates with other public figures of the day. He regularly attends performances at the Globe Theatre, for instance, and another diary of his, the *Book of Plaies and Notes thereof* from 1611 contains first-hand accounts of performances of several of Shakespeare's plays, including a review of the appearance of Banquo's ghost in *Macbeth* and a lively response to Lady Macbeth's sleepwalking. The medical establishment are after Forman — they think he's a quack — but nevertheless he has a reputation for curing chronic conditions so people mostly bring health problems to him.

A portrait of the astrologer that he commissioned himself hangs on the wall behind his desk. The shelves groan with vellum-bound astrological compendia. In front of Forman is an 'ephermeris', a table of planetary motions to which he can refer in order to calculate the positions of the planets at any given time. Forman holds up his pen ready to hear Burton's problem.

According to the casebook, Forman asks a series of standard questions of the young man consulting him: name, age, complaint, etc.[25] Burton reports a general malaise and stomach complaints. He wants something to help with his low mood. Perhaps he already thinks it may be the affliction he has heard so much about: 'melancholia'.

Forman consults the table of planetary positions, looking for the date and time of the consultation, then he plots this onto a chart of the heavens, divided into the twelve signs of the zodiac, or 'houses'. The horoscope is calculated by the position of the known planets – Mercury, Venus, Mars, Jupiter, Saturn – the sun and the moon, in relation to the twelve astral sections. There are different kinds of horoscope. A 'nativity' maps the positions of the planets' time of birth. It might be calculated retrospectively based on events in a person's life. Other horoscopes are drawn up in answer to the question of when it would be a good time to do something, or a horary question with a simple 'yes' or 'no' answer, but most commonly they are calculated based on the alignments at the time of asking the question. The position of the stars and planets at first consultation is considered important. This is Burton's first visit and he is here to ask about his health.

Now Forman reads the horoscope, based on a series of rules about position of planets in houses. Each house represents a particular set of categories: for example, a principle such as 'life or death' (eighth house), a concept such as 'home' (fourth house) or a relationship (seventh house). The first, sixth and eight houses are the most important in relation to questions of medicine, and the attendance of Saturn or Mars in these houses was considered ominous.

Forman then consults the many volumes of rules which line his shelves. Thousands of configurations of star–planet alignment are possible.

Forman diagnoses melancholy, and prescribes various medicines and purges for his symptoms. Burton listens to the diagnosis and then, just as he's about to go, Forman throws in another piece of information. The student nods, pays and leaves the Stone House, stumbling out into the street. He takes a coach back to Oxford.

In his popular guide to astrology, Forman stressed the importance of the relationship between the physician and the patient, noting that trust could be as important as any treatment prescribed. His consultations were prescient of the 'talking cures' of psychotherapy centuries later. The bundles of casebooks reveal that many people come to him for advice on relationship, and even on sex.

By choosing to stay in London during outbreaks of plague in the early 1590s Forman earned himself a public reputation as a

courageous man. His medical successes also attracted the College
of Physicians, but not in a good way. They consider him a fraud,
and he in turn believed they were persecuting him, the belief
escalating to such a degree that he told friends they intended
to kill him. He came from a Wiltshire family which had been
stripped of its entitlement by the dissolution of the monasteries
in the previous generation. He was a self-made man, by way of
work as a hosier and grocer's apprentice.

Before the next client arrives, he writes up the notes from the
last consultation.

He begins by listing the patient's symptoms: 'grife & coulde a
shouting in his veins/pain had belly. a[nd] the blod is stopt & hath
not his course. A gret heavines and drowsines in the hed...moch
wind in his bowels...he will have a palsey or a quatrain fever.'

He then records what he's just told the young man: 'or yt it is
of melancholy prepare 3 dais and purge he carrieth death upon
him...& will die suddenly'.[26]

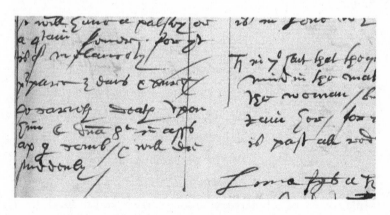

Simon Forman casebook, 28 June 1597. 'Robart
Barton of 20 Yeres...will die suddenly.'

At 5.30 p.m. on 25 June, Burton returns to Forman for a second consultation about the same symptoms: 'Robarte Burton...second questo, yt is in his hed noz & belly he took 1 of pills.'[27]

Three further visits are recorded: 'Robart Burton of 20 yeres 1597 the 21 July...diz himself...moch pain hed & moch wind and melancoly remn roning to the reins & belly & stopping the vains which will cause a fever & vexation of the mind some 17 days hence/...give him a dietary purge for 4 dais/RC 2 he must have yt for 4 dais and pay a noble';[28] 'Robarte Burton of 22 yeres 1597 the 19 Aug pm...moch stopping in the stomacke and awind in the belly...he is heavy & unlusti and yt pricks in his hed';[29] 'Robart Burton of 20 yeres the 11 of Octob am... A burning in his hands & knes & a wind in the belly.'[30]

Burton returned again and again. He'd been told, after careful calculation, that he 'carrieth death upon him' and would 'die suddenly'. But he still went back for more. Forman offered him something with the horoscope that, in his melancholic state, he found useful.

Burton is erratic in the *Anatomy*; his tone jumps around, his sense of purpose shifting suddenly from light entertainment to a sober warning. He talks about a fearful illness that will creep up on you 'so by little and little, by that shoe-ing horn of idleness, and voluntary solitariness, melancholy this feral fiend is drawn on', leaving a 'cankered soul macerated with cares and discontents' and, later still, 'agony'.[31] The next minute he raises a sardonic eyebrow at the pointlessness of human activity, particularly the futile misery of scholars trying for years on end to finish their sprawling books. The publishers demanded English but Burton was much more comfortable with Latin. He was well acquainted with the word 'delusionem', which brought us the word 'delusion' in the sense of mental derangement meaning 'a deceiving': the act of misleading someone, deceit. It contains within it the verb 'ludere': literally 'to play'. There is in the very word itself a sense of leading astray, of a twinkle in the eye. Delusions are a guessing game, but very serious messages are smuggled out of something which feels like a sport.

For fifteen years after his first consultation with Forman, Burton lives and works beneath the reference points in the sky that tell him when he should expect this sudden death. It's easy to imagine how a prediction like that from a charismatic figure like Simon Forman could cast a shadow over a young life, especially over one already menaced by depression.

Around 1603 Burton draws up his own nativity, in his note-book, using Forman's guide to astrology for direction. He captures a snapshot of the position of the twelve signs of the zodiac in the night sky at the time of his birth. The 'sign' rising in the sky became his first house, and the other eleven houses were allocated anti-clockwise, across an ellipsis, relative to that sign.

He consults his ephermeris tables to establish the position of the planets, the sun and the moon, at the time of his birth, and determines which houses they were sitting in. Each house with a theme: 'health' or 'communication' or 'home', for example. Then he starts to analyse the lines which crisscross the chart, showing how the spheres sit in relation to each other. He begins to tease out the subtler meanings. He returns to Forman's guide for help with the interpretation. The tables tell him when Saturn or Mercury – the most ominous planets – will be in his eighth house. This house relates to matters of life and death. He shouldn't look, but...

There it is. The planets will cross his eighth house, and signpost a climacteric event. In the year he is sixty-three.

The belief that he will 'die suddenly' now has a date in the calendar. Here is a timeframe that he must work within.

No wonder Burton returned to the literature of the classical world, where supernatural forces had the upper hand and the gods played puppet masters to their mortal pawns' delusional acts. The poets and playwrights gave expression to man's powerlessness in the order of things. In this world, delusion is a seductive force. The concept is embodied in classical mythology by a vivacious young female in the pantheon of the gods of ancient Greece. 'Ate', daughter of Eris and Zeus, is the personified spirit of delusion and blind folly. In Homer's *Iliad*, a work of the eighth century BCE, Agamemnon addresses Achilles: 'Delusion [Ate] is the elder daughter of Zeus, the accursed who deludes all; her feet are delicate and they step not on the firm earth, but she walks the air above men's heads and leads them astray. She has entangled others before me. Yes, for once even Zeus was deluded, though men say he is the highest one of Gods and mortals.'[32]

In Apollonius Rhodius's Hellenic epic 'Argonautica', Hera tells Thetis: 'Even the Gods are sometimes visited by Ate.'[33]

Nonnus's fifth-century poem, 'Dionysiaca', features Hera, who in her vendetta against Dionysus sends Ate (delusion) to persuade the god's young lover Ampelos to ride on the back of a bull. In so doing, she plans to bring about his death: 'Ate, the deathbringing spirit of delusion, saw the bold youth straying on the mountain away from Lyaios during the hunt; and taking the charming form of one of his agemate boys, she addressed Ampelos with a coaxing deceitful speech…'[34] In *Agamemnon*, Aeschylus refers to delusion when he describes the state of the king's mind after he sacrifices his daughter: 'But when he put on the harness of Necessity, breathing an impious, unholy, unsanctified turning of mind, then he changed to a dare-anything purpose. For delusion, evil-counselling, wretched source of evils emboldens mortals. So he dared to become a sacrificer of a daughter, and arousal of woman-reprising war, the first offerings for his ships.'[35]

The golden age in Athens also saw the inner landscape of the 'psyche' being mapped for the first time in the plays of Aeschylus, Sophocles and Euripides, a landscape and cast of characters Freud returned to. Burton brings into his collection examples of the world's great love stories from literature that look forward to a modern approach to psychological problems. He was applauded centuries later for being ahead of the game by Dr Johnson, Laurence Sterne and the Romantic poets.

But try as he might, he can't contain his own melancholy. When the time comes to appraise his own career, we don't hear a stoical scholar, we hear a frustrated man: 'I am not poor, I am not rich; I have little, I want nothing: all my treasure is in Minerva's Tower. Preferment I could never get, although my friend's providence, care, alacritie was never wanting to doe me good, yet either through mine owne default, infelicity, want or neglect of opportunity, or iniquite of times, preposterous proceeding, mine hopes were still frustrate, and I left behind, as a dolphin on shore, as Diogenes to his tubbe.'[36]

He likens himself to the down-and-out Greek philosopher famous for sleeping on the streets of Athens in a broken barrel, overlooked for promotion. He is resentful, critical of himself, maybe a little paranoid, though he says he's resigned to his lost hopes. In his sketch of daily life earlier in his career he portrayed himself as 'sequestered from those tumults and troubles of the world', buried in his work at Christ Church in splendid isolation as he had spent his childhood in Lindley Hall encircled by the moat. He can't hide from the tumults and troubles in his own mind. The melancholia won't leave him alone.

The death of Simon Forman is documented by William Lilly, another astrologer. It's Lilly who will predict the Great Fire of London fifteen years before it happens in a woodcut print portraying his vision of houses and people engulfed in hellish talons of flame.

Lilly slips the story of Forman's demise into an account of his own life and times. He begins with the gossip – a report that one Sunday afternoon in September 1611, when Forman was fifty-two and living in Lambeth to escape the jurisdiction of the city, he told his wife that he would die the following Thursday night.

> [M]onday came, all was well. Tuesday came, he was not sick. Wednesday came, and still he was well: with which his impertinent wife did much twit him in the teeth. Thursday came, and dinner was ended, he very well. He went down to the waterside, and took a pair of oars to go to some buildings he was in-hand with in Puddle dock. Being in the middle of the Thames, he presently fell down, only saying, 'Am impost, an impost,' and so died.[37]

Whatever the truth about any events or conversations in the run-up, Forman was dead. Did news of his death make its way to Burton in Oxford? Did he hear the reports that Forman had predicted the precise time it would occur? Burton is thirty-five years old in 1611. He has been researching melancholy for at least ten years, and living with waves of this debilitating, depressive illness. If Forman could predict his own death, it stood to reason he could predict it for others. He has his horoscope as further evidence. Burton throws himself into the project of containing the subject inside the pages of his book, pushing ahead with the organisational work after a decade of collecting scraps. He publishes the *Anatomy* ten years later when he is forty-five. The book now stretches to three volumes. There are modifications to the stream-of-consciousness with each subsequent edition reflecting changes in his thinking. He adds material over many years, setting a rolling reading exercise for his audience and offering a practical cure for melancholy for anyone who attempts to keep up with the project. And many did. The *Anatomy* ran to four further editions by the time of his death, Burton stuffing new case studies, symptoms and cures into each, followed by a sixth posthumous edition in 1652 when it had overflowed again into half a million words. The market was hungry for it and he was paid by Henry Cripps, the publisher, to do the rewrites. Burton's last will and testament was written just five months before he died when he was by all accounts fit and well. It shows that *The Anatomy of Melancholy* had made him rich. It found its way onto the influential theologian Richard Holdsworth's seventeenth-century Cambridge reading list for young gentlemen. The work has orbited the world ever since.

How successful it's been in helping others wrestle their melancholy into submission is hard to judge, but Burton certainly never got the upper hand over his own. On his memorial Burton gives the last word on his life to melancholy. He organised melancholy and melancholy organised him. The circumstances of Burton's death are not so clear, mired as they are in centuries of gossip. Burton may not have killed himself at all. The rumour may simply have been Chinese whispers at the university. The fact that Burton was buried in the cathedral makes suicide less likely – it wouldn't have been allowed – but he does seem to allude to the sin on his memorial there when he says in the epitaph that melancholy gave him death. Likewise, we only have William Lilly's word for the story around Forman's death. It is all hearsay.

There's an anecdote about Burton which gives a clear sense of the tightening grip melancholy had him in. Unable to free himself, and on the rare occasions that he could drag himself away from his work, he apparently had a habit of setting off down to the river to watch the lively goings-on where the barges were moored. A Mr Granger is quoted in the 'Notes on the Author' in William Tegg's 1863 edition of the *Anatomy*. He says that Burton 'composed this book with a view of relieving his own melancholy, but increased it to such a degree, that nothing could make him laugh, but going to the bridge-foot and hearing the ribaldry of the bargemen which rarely failed to throw him into a violent fit of laughter'. Burton imitates the laughter of his alter ego Democritus in the face of the follies of mankind. But with only one thing left in the world that could coax a smile out of him, the cynical detachment is unconvincing and the laughter sounds hollow.

His horoscope is useful. A sea of melancholy surrounds him. Here is definition, measurement, logic. Burton wasn't so much 'deluded with future events' as hanging onto them as he might cling to a buoy during a storm. The fixed points of an astral chart serve a life-saving purpose in that context. They are something by which to orientate himself. Astrological reference points orbited his capacious mind for most of his adult life, promising to steer him.

Burton tries to hide behind a satirical literary persona. He seeks sanctuary in scholarship in his beloved Christ Church. A few brief notes from an astrologer's casebook reveal snapshots of a vulnerable young psyche. A man who wants to make sense of a depressive illness. Maybe these notes are the truest voice of Burton's that we hear.

Ars Moriendi, 'Temptation through despair',
Master E. S., *fl.* 1450–1467.

CHAPTER IV

Francis Spira and the 'Delusion of Despair'

May 1548, Venice, northern Italy. A forty-six-year-old lawyer, Francis Spira (born Spiera) stands before the Legate of Venice. He is on trial for the heresy of his Protestant faith, and he must recant or leave Italy. He stands in his plain clothes under the scrutiny of the judges: a cloth cap and doublet over a simple tunic, a neat beard, eyes appealing for understanding. He finds none. Spira cannot win. Despite the immense pressure to renounce his conversion, he believes in Luther's ideas. He is also in mortal fear of what will become of him and his family if he does not renounce his faith. Caught between the religious authorities and the demands of his god, he has no discernible way out. He collapses. After coming to, and making a final assessment of his options, he agrees to recant all his Protestant beliefs, in writing, and he signs the necessary documentation.

Spira returns to his hometown of Cittadella, near Padua. Until recently, he has been a man of high standing. The reception is muted. People are wary. Neighbours, some of whom have denounced him to the authorities – he doesn't know who, but he has his suspicions – don't meet his eye. He has administered the law to the community for so many years but goes to bed expecting that it will be used to punish him. He is also tortured with shame about his public denial of his true belief. His eleven children are asleep in the house around him. He is desperate to think of a way to protect them. He can't sleep either. After a night spent tossing and turning, he comes to a decision. He will do the everything necessary to signal that he adheres to the terms of his freedom. He pays a fine, even makes a shrine for the Eucharist. He then goes in front of a congregation of two thousand and reads out the document he signed in Venice word by word.

But on the way back to his house and the family whose immediate safety he just secured God appears to him. In the words of Nathaniel Bacon, a Puritan lawyer and politician who publishes the most sensational account of Spira's story, and makes it famous, God rebukes him in the sternest terms: 'Dost thou well in preferring wife and children before Christ? Is the windy applause of the people, better indeed than the glory of God?'[1] The sufferings of the present life, he warns, will be nothing compared to the glories that would be revealed by doing the right thing and returning to the true path.

Confronted with the terror of God, Spira is frozen to the spot. How is he supposed to weigh this future pain against the present dangers of the world that threaten his family? He has already betrayed his Protestant faith, and if he goes back to Luther the worldly punishments will be unimaginably severe.

He reaches his house, but as soon as he gets there the voice of God assails him again: 'Thou wicked wretch, thou hast denied me, thou hast renounced the covenant of thy obedience, thou hast broken thy vow, hence Apostate, bear with thee the sentence of thy eternal damnation.' In Bacon's version, the conversation reads like a play or a witness statement in a police report, and it's easy to see how this account, with its lifelike high drama, became so popular.

The impact on Spira is dramatic. The words bring about the climax of the physical and mental crisis that's been gripping him: 'he trembling and quaking in body and mind; fell down in a swounde; reliefe was at hand for the body, but from that time forwards he never found any peace or ease of mind...he heard continually that feareful sentence of Christ that just Judge: that he knew he was utterly undone...others not looking so high as the Judgement of God, laid all the blame upon his Melancholicke constitution; that over-shadowing his Judgement, wrought in him a kind of madness.'

Some, Bacon says, took one look at the situation and ascribed Spira's state to that depressive mental disturbance which had swept across Europe, affecting scholars like Spira and distorting their imaginations. Spira's version of melancholy was not the man reclining by a meandering stream, with a thoughtful middle-distance stare out into nature. His swoon marked the beginning of a strange decline in his physical health, as well as a mental unravelling.

Spira experiences an epiphany – he is damned – and the story of the revelation travels at lightning speed across Europe. Spira's name is anglicised on the way.

Some fifty years after the events unfold, it will eventually reach a man who is collecting stories of 'Melancholicke' constitutions just like Spira's, a certain Robert Burton. The story catches his attention. He re-spins it in *The Anatomy of Melancholy*: 'There is a most memorable example of Francis Spira, an advocate of Padua, Ann. 1545, that being desperate, by no councel of learned men could be comforted; he felt (as he said) the pains of hell in his soul; in all other things he discoursed aright, but in this most mad.'[2] Burton had another fabulous case of delusion to cram into his book. As he observed, and understood only too well, believing that you are doomed can by itself bring the worst fears to fruition even in the face of plenty of evidence to the contrary.

Back in 1548, on 27 December, only a few months after his collapse and still in Cittadella, Spira is at the end of eight weeks of self-starvation. Around his bed are several men – a group comprised of other lawyers, doctors and clergy all intent on helping him. He is skin and bone, but refuses all attempts by these friends to get food or liquid into him. They are begging him to eat but he is so sure of his own damnation that he is continues to hold out, and does not respond to any of the many cures his friends attempt. According to Bacon's version, Spira's understanding is still 'active'. He is 'quick of apprehension, witty in discourse' but at the same time rock-solid in his conviction that he is condemned to hellfire.

His friends take him to physicians at the University of Padua in search of a cure for this 'Melancholicke' constitution: 'They could not discern that his body was afflicted with any danger or distemper originally from itself, by reason of the over-ruling of any humour; but that this Maladie of his did arise from some griefe, some passion of the minde, which being overburthened, did so oppress the spirits, as they wanting free passage, stirred up many ill humours, and these ascending up into the brain, troubled the fancie, shadowed the seat of the judgement, and so corrupted it. This was the state of his disease, that outward part that was visible to the eye of nature; this they endeavoured to reform by purgation…but all their skill effected nothing.'[3]

These friends cannot confidently blame Spira's problems on a humoral imbalance as they once might have. They cast around for the ingredients behind it, identifying in the recipe both spiritual and physical cause, and a dash of psychological. Ultimately the best explanation they can come up with is a kind of mental overload. Spira thinks they have all missed the point entirely. He chastises the people who are trying to help him: they must understand that only God can be his physician. This is not a disease that can be cured with potions and purges. 'No persuasion could ease him,' writes Burton in his pass at the story. 'Never pleaded any man so well for himself, as this man did against himself.'

Spira does indeed die, later that day, of despair. Some say he starved to death, and rumours suggested that he had done this deliberately – committed suicide – which was, of course, an abominable sin.

So the story goes, anyway. The intriguing band of men around Spira's bed comprised the following: Pier Paolo Vergerio, Bishop of Capodistria, who was himself under investigation by the Inquisition regarding his religious position; Matteo Gribaldi, professor of civil law at Padua and colleague of fellow lawyer Spira, Henry Scrymgeour, a Scottish Protestant from St Andrews, referred to in the most popular retelling of Spira's story simply as 'a Scot'; and Sigismund Gelous, 'a Transylvanian'.[4] The group are representatives of the two most powerful agencies in the city of Padua at the time: the law and the church.

This death scene is quintessentially medieval. *Ars Moriendi*, literally 'the art of dying', a practical guide to dying, was selling fast across Europe at the time. The book counselled on what to do when facing your demise. This was one of the first texts widely and quickly distributed throughout Europe by the movable-type printing presses pioneered by Johannes Gutenberg. One particular chapter advised against the temptation to despair when death was close. A pair of accompanying illustrations highlight the jeopardy. Gargoyle devils and assorted grotesques carrying knives will prance around the bed if you give in to it. If you stand firm and keep the faith, angels will attend you. A horse might even lie down in tribute at the foot of the bed.

Ars Moriendi, 'Consolation through confidence', Master E. S., *fl.* 1450–1467, Ashmolean Museum, University of Oxford. See frontispiece 'Temptation through despair'.

In Matteo Gribaldi's version of the story, which was probably the first published account, Spira experiences such agonies for denying his Protestant conversion that he sees devils sticking pins in his pillow, and a fly on the wall becomes a visiting demon.[5] These images are straight out of the *Ars Moriendi* playbook. The Protestant and medieval worlds are fighting to the death in the body of Spira.

It turns out that several of those who witness his death are committing the words and actions of this painful crisis to memory and each will pen his own eyewitness version of events. These rival versions will then make their way from northern Italy to the rest of Europe, each in its own way amplifying the story. It's not the breakdown of a provincial Italian lawyer any more; it's a blazing exemplar of what lies in store for any individual if he rejects the true religion for a heresy. The tectonic shifts of the Reformation shake the whole of Europe and everyone has to jump one way or the other – stay on the side of Rome, or make the leap over to Martin Luther and John Calvin.

Spira's home of Cittadella was a fortified 'little city', an outpost of Padua, under the jurisdiction of Venice; its defensive stone walls, laid in a polygonal shape as a first line of defence which came in useful during numerous regional wars over centuries, and its thirty-two towers and gates would have given it passing resemblance to Burton's Oxford. Cittadella stood at the border with more northerly Europe, across which the Protestant literature travelled into Italy. And some of the glass from Murano set off for new homes.

Spira (the stories all agree on this) was a successful figure in Paduan society, worldly and well educated, when he first came across the writing of Martin Luther. He entertained these new ideas cautiously at first, but consumed more and more, until, slowly and inexorably, they convinced him. Six years after his conversion, he was sharing these Lutheran ideas at public meetings around his district. He was too persuasive for his own good, it seems, and, in November 1547 his neighbours reported him to the Inquisition. Venice and its population had grown cautious. The region was caught on the northern fringes of the country where the Catholic Church was particularly susceptible to a leaky border which allowed heretical ideas in. Rome was clamping down on these threats to papal and imperial powers.

The optics on Spira were not good. A book from the blacklist had made its way into his hands. When his trial began in Venice in May 1548, he admitted seeing the *Beneficio di Cristo*, a publishing sensation of the Italian Reformation. The last chapters were distinctly Calvinist and Predestinarian. His guilt was obvious.

A Reformation historian and editor, Caelio Secundo Curione, swiftly gathers together several alternative accounts of Spira's suffering including Gribaldi's, the very first one to make it out to England. Curione is a man with disruptive instincts and he weaponises the collection in a single volume in Latin, in 1550, with a preface by Calvin himself. The story circulates throughout the continent, surfacing widely in sermons and treatises. The same printing press which was so instrumental in the initial spread of Protestant ideas also disseminated Spira's story across Europe. It made its way to England, where Edward VI was on the throne following his father Henry VIII's death. No one knows yet that his sister Mary will return the country to Catholicism in a just few years.

An Italian Protestant exile, Peter Martyr Vermigli, was responsible for bringing the story to Oxford. Vermigli had come in 1549 at the invitation of Archbishop Cranmer during Edward VI's reign and retold the story of Spira as part of his lectures on Romans only a year after Spira's death.[6] These lectures were written up and the cautionary tale of Spira reached Robert Burton, half a century after the events took place.

We can imagine Burton getting himself a copy of Curione's confection of stories to arm himself with more takes on what happened to Spira and holing up at his desk in his chilly Oxford rooms to consume them under a guttering candle. The lectures connected what was, on the face of it, a very Italian story to Burton's day-to-day life in Oxford as well as to the doctrinal debates raging in his city's own churches. A strange tale from faraway Italy was immediate and germane to his experiences. Spira caught Burton's imagination. The story was from a previous generation, and the volume was turned up on the religious content, but it mirrored something of Burton's persistent sense of foreboding.

Publicly Burton took an aloof position on Spira. He reglossed the story and wrote about him as an example of 'religious melancholy', and an example of a delusion. Spira was experiencing a mental illness, in other words, with religious content. Burton calls him 'most mad' and diagnoses Spira's religiosity. Spira is not a prophet but a madman; despair is not a spiritual failing but a disease. To non-believers, Spira's pain is an agony of alienation

and negation. Privately, Burton was not so rational. At the time
he first read the story he was fixated on his horoscope. In print he
cautioned others to remember that simply believing a prediction
could make it come true, but he understood perfectly well how a
sense of impending doom could linger regardless.

A belief that an eternal punishment awaits has driven many
memorable characters of page and stage from the past two millennia to ruin, the standard-bearer being Christopher Marlowe's
tortured creation:

> *Ah Faustus.*
> *Now hast thou but one bare hour to live,*
> *and then thou must be damned perpetually…*
> *The stars move still, time runs, the clock will strike.*
> *The devil will come, and Faustus must be damned…*
> *Cursed be the parents that engendered Me:*
> *No, Faustus, curse thyself, curse Lucifer.*
> *That hath deprived thee of the joys of heaven.*
>
> Christopher Marlowe, *Dr Faustus*, c.1592

When Marlowe sat down to write *Dr Faustus* in around 1592, the
same man was playing on his mind as had played on Burton's:
Francis Spira. This man caught in the crossfire of sectarianism,
who believed he was damned by God, continued to haunt the
sixteenth and seventeenth centuries.

The most widely disseminated English account of Spira
over the ensuing years by far was Nathaniel Bacon's version of
1638. Bacon was a prominent legal brain who sat in the House
of Commons and *A Relation of the Fearefull Estate of Francis Spira*

was a spin on the original Latin records of the case. An instant sensation, its sparkling dialogue between God and the apostate captivated readers. The manuscript was a talking point before it was even published thanks to the careful work of a London turner, Nehemiah Wallington, who had copied the whole book out, word for word, in 1535.[7] Like the *Anatomy* its appeal proved enduring and it was reissued many times, ten times in England and eight times in the American colonies, all the way through to the eighteenth century.

It's likely that Marlowe, like Burton, read many different versions of Spira over the years, but Bacon's portrait of 'a reprobate like Judas or Cain, who casting away all hope of mercy, fell into despair' had an over-the-top energy that trumped the other iterations.

Within a decade of his demise, Spira's life and death had acquired a new dimension of meaning. He was now an archetype, a symbol of the idea that suffering was preferable to damnation and the man himself stepped out of sight. Spira's story was covered in theological tracts, sermons, plays, ballads and 'wonder books' for the decades which followed. Even the Protestant queen Lady Jane Grey cited him in a letter warning her father's chaplain, a Mr Harding, about what his conversion to Catholicism would mean for him under Queen Mary. She reminds him of the 'lamentable case of Francis Spira, whose case (me thynke) should be yet so greene in your remembrance, that being a thing of our time, you should fear the like inconvenience, seeing you are fallen into the like offence'.[8] Don't get caught out, like Spira, she says. She presumes that Harding knows the story well. John Bunyan certainly did, recalling in 1666 how he 'did light upon that dreadful story of that miserable mortal, Francis Spira; a book that was to my troubled spirit as salt, when rubbed into a fresh wound; every sentence of that book, every groan of that man, with all the rest of his dolors, as his tears, his prayers, his gnashing of teeth, his wringing of hands, his twining and twisting, languishing and pining away under than mighty hand of God that was upon him, was as knives and daggers in my soul.'[9] Bunyan had suffered a similar torment, and wanted a pardon that did not come.

We might brush the story off now as a maxim for all times about how one man's delusion is another man's religious belief, or how 'true religion' can't be easily told from 'heresy' since the judgement is entirely a matter of faith.

But Spira's delusion came out of the schisms of the Reformation, and his agony tells us in specific terms about the prevailing anxieties in continental Europe at that time, where people did have faith, in one side or the other, and believed in the reality of damnation and the endless torments that would beset a heretic, or a sinner who fell into despair. There were no inverted commas around any of these concepts.

Among English Puritans, the temptation to despair was the most feared of Satan's invitations because it was the loss of hope in one's own salvation. Spira's story was held up by clergy as a warning specifically of this sin of talking your own life. A woodcut print appeared featuring the progress of Spira, alongside another famous despairer, the Englishman John Child, a teacher who had committed 'unnatural murther upon his own person' in Brick Lane, Spitalfields, in east London in 1684.

There continued to be a high turnover of religious regimes over the next century, and Spira's predicament remained relevant to a wide spectrum of the society who shared the challenge of adapting. It's no surprise that Spira inspired writers like Marlowe with keen sensitivity to any fault lines in society. It's not surprising either that Spira's story became a powerful Protestant propaganda tool, brandished by the authorities to keep believers on the straight and narrow, for decades to come. The authors of the primary Spira sources weren't dispassionate observers either. Pier Paolo Vergerio, Matteo Gribaldi, Henry Scrymgeour and Sigismund Gelous were all Protestant converts and navigating the political and theological chaos themselves. Their accounts start to look a bit like propaganda, too.[10]

Catholics had a use for Spira as well. They were stretching the point a bit, but his torment was worked into an example of the dangers of Lutheranism. Puritans sympathised with Spira, seeing his inner turmoil before the despair as simply an extreme example of the experience of all godly Christians.

Religious switchbacks continued into the eighteenth century and people still had to adapt to thrive though the skills required could be represented in less solemn terms. The popular satirical song 'The Vicar of Bray' charted the career of a Berkshire clergyman as he contorted his principles in a brazen attempt to keep his job under a succession of capricious kings and queens. Spira's story is at its heart a painful one and more relatable at the time than a finger-wagging cautionary tale. On one level this was a personal, existential crisis. It wasn't heresy, it was delusional thinking that allowed Spira to bow out of an impossible situation. Spira's delusion was so extreme that it swiftly led to his mental and physical breakdown, but anyone living through the sixteenth and seventeenth centuries at the time would have recognised a desire to withdraw from the world. A world where you could never make yourself safe however hard you tried.

Can we understand more about Spira, the man and his delusion, behind the cipher he became for the great theological fights of his age? As usual, Bacon's popular work includes the most colourful and up-close detail on his idiosyncrasies. He describes Spira as: 'an advocate of great rank and esteem, being of known learning and eloquence, of great experience, of carriage circumspect and severe, his speech grave and composed, his countenance sharp and austere, every way befitting that authority whereunto he was advanced, endowed with outward blessings, of Wife and eleven children, and wealth in abundance'.

This lively description really points up the remaining puzzles about Spira. In simple terms the delusion of despair offered him a way to surrender, but it doesn't complete the picture. His predicament wasn't unusual. At the height of Spira's troubles, a major moment in the Counter-Reformation is kicking off. This is a rebuttal to Protestant power. The Council of Trent convenes about a hundred miles north of Padua. The assembly continues in stages from 1548, Spira's critical year, until 1563. It introduces sweeping decrees, nailing down all the dogmatic teachings of the church on purgatory, heresy, the lot. Spira's crisis coincides with a Catholic Church backlash but most people of his standing were dealing with the exact same issues as he was. Everyone had to negotiate the doctrinal fog, as well as clampdowns, and cardinals and lawyers, just like the general population of Venice, Padua and Cittadella contradicted themselves from time to time over the years, somehow groping their way through. Many of Spira's peers had not yet made their minds up on matters of religion.

People had faith but they also lived with contradictions, as you might live in a new house furnished with a grandmother's old dresser or bed. Whether out of habit or nostalgia, the old wasn't just thrown out. So what pushed Spira to the extreme?

He was a rich, well-respected, fiercely intelligent and apparently mentally balanced family man, but his actions sabotaged a solid good fortune and social position, and his life. Later on, his belief that he was damned stood up to all the reasonable arguments of his peers in public. He was a lawyer and interrogating a position with logic was what he did.

Spira trained for at least seven years at the distinguished University of Padua to achieve his credentials as a doctor in law. At the time of his crisis he had been practising for several decades. Law was for the gentry and he would probably have come into the profession as the next in line of several generations; law was not something you could move into easily. Padua was a litigious world. Lawyers were in high demand, and being a lawyer brought you power, influence and money. Advocates like Spira offered many people their only access to the written word, and were also the only route to legitimacy for social and financial transactions or disputes.

For as long as anyone could remember, lawyers had been called into one particular type of dispute above all others: 'matrimony'. All it took to be married at the time was an exchange of consent between the parties in real time. You didn't need a priest or witnesses. When relationships broke down, the resulting accusations quickly became a matter of 'he said, she said'. A person couldn't be guilty of adultery if they were never married in the first place, so the legal pressure was on. Telling the injured party from the guilty party was all but impossible if no one knew for sure whether such an 'exchange of consent' had taken place in the first place. Reputations were on the line, prison sentences loomed large and you needed the most silver-tongued advocate on your side. You would pay top rate for services if you could afford it. It was your only chance at agency in these situations: it would be worth the fee. This was the fuel for protracted legal theatre, featuring the most persuasive advocacy available. If argument failed to influence the outcome, bribes were passed under the table. Civil lawyers like Spira, straight as a die according to reports, as well as notaries (legal accountants) were in the money. Men in these positions often invested in land and property both inside and outside of town, and accumulated a considerable stash by drawing incomes from tenants. They found themselves at the centre of a complicated, and lucrative, legal, financial and emotional nexus within their community.

Venetian government records related to the Inquisition of Francis Spira contain transcriptions of the pre-trial interviews of Spira after his arrest, as well as of others deposed in relation to his case. These documents suggest that, initially at least, the conflict between Spira and certain members of his own community was not related to his conversion at all.

During the course of the interview, Spira makes reference to his associates' decision to turn him in to the Venetian authorities. He suggests that there was another motivating factor for this – the marital problems of a female member of his family. By the time of his trial in June 1548 Spira had eight children at home, two married children living away and a pregnant wife. He states that the persecution 'was brought about by my wish to preserve my honour and that of one of my daughters – an affair which was notorious'.[11]

There are no further details of this 'notorious' affair. Was it a case of adultery, and if so, who had committed it? Could questions have been raised about the marriage's legitimacy? It begins to look as though his professional specialism had come a bit close to home. This was a society which operated within the strictest codes, and they were about to get stricter after the Council of Trent. Cittadella was also a place where gossip and rumour could dole out one of the worst punishments for a transgression: shame. A lawyer like Spira would already have a fine-tuned sense of shame, and its power to strip a family of status and influence and ostracise it. Family honour was the currency in a town like Spira's, and without it access to good society was denied. The head of any family would want this access for his household. It would have been a major blow when the tables were turned on him and he found his own daughter involved in a scandal.

Another document in the records of the Venetian Inquisition suggests that there were other uncomfortable thoughts occupying Spira's mind. Guilty thoughts, that had nothing to do with his conversions or renunciations, or his daughter's notorious affair. They were connected to his past actions in a professional capacity and things he shouldn't have done.

After the trial Spira had stayed with his cousin, Nardini, in Padua. Nardini's testimony was given to the Inquisition about two weeks after Spira's death. In it, Nardini said that Spira was mad, a common defence of families when a loved one was facing the Inquisition because it might get them off the harshest penalties. But he also implied that Spira needed a pardon from God for the way he had made his money when practising as a lawyer.[12]

In Bacon's version of the story, a by now unravelled Spira backs this up. He confesses: 'I was (saith he) excessively covetous of Money, and accordingly applied myself to get by Injustice, corrupting Justice by deceit, inventing tricks to delude Justice; good causes I either defended deceitfully, or sold them to the adversary perfidiously; ill causes I maintained with all my might… Thus having worn out forty four years, or thereabouts, and the news of the new, or rather, newly revived Opinions of *Luther* coming into those parts.'

Spira could no longer keep a lid on his confessions. He admits he had a slavish devotion to moneymaking. If he's telling the truth about what he got up to, he might well feel remorse. There's no obvious reason for him to lie about selling people out, left and right, in order to line his own pockets. The admissions are detailed: he'd lied in court about innocent clients and protected criminals. He'd flagrantly perverted the course of justice, defiling the noble institutions of the law. He'd kept it up for years. Significantly, it seems he had managed to keep his outrageous corruption a secret, or at least avoided exposure, maybe through blackmail or bribery, well into middle age. Acquiring wealth through the accruing of interest was usury – morally suspect in Europe, even sinful. This man who'd been piling up ill-gotten gains for decades knew his punishment was also accruing and would be waiting to be caught. No wonder Spira was unstable.

Burton tried to make sense of a delusion like Spira's. In the *Anatomy* he makes room for a whole chapter on 'religious melancholy' and includes a few more tormented souls with delusions of damnation.[13] He refers in passing to a 'Cardinal Crescence [who] died so likewise desperate at Verona', and other cases he has heard of where a person's belief took them to the point of suicide. He ponders on whether this is a failure of reason or imagination.

In his chapter on 'Causes of Melancholy' Burton blames Spira's distress on the scourge of human curiosity about the future. We can't possibly know what lies in store for us after death, but we can't help probing:

> To these tortures of fear and sorrow may well be annexed curiosity, that irksome, that tyrannizing care, *nimia sollicitudo*, 'superfluous industry about unprofitable things and their qualities,' as Thomas defines it: an itching humour or a kind of longing to see that which is not to be seen, to do that which ought not to be done, to know that secret which should not be known, to eat of the forbidden fruit. We commonly molest and tire ourselves about things unfit and unnecessary, as Martha troubled herself to little purpose. Be it in religion, humanity, magic, philosophy, policy, any action or study, 'tis a needless trouble, a mere torment. For what else is school divinity? How much doth it puzzle! What fruitless questions about the Trinity, resurrection, election, predestination, reprobation, hell-fire, etc., how many shall be saved, damned?[14]

Burton shakes his head sadly at the futility of it all. No one can resist taking the bait on what the future holds, certainly not him.

Spira's lens on the world was filtered by the Protestant idea of election. This compels everyone to search for clues as to whether or not they have been chosen by the Grace of God and saved, or damned, but no one will find a conclusive answer until Judgement Day. Spira's religiosity sends him searching for evidence of his damnation. He is persecuted by God, chased and harangued by him in the street, and his delusional state starts to look a lot like what we think of now as 'paranoia'. He is aware of powerful forces at work, convinced that there is a Grand Plan, and also that he is not in control or in the know. Nevertheless, the schema gives Spira reference points to orientate himself by, like Burton's maps of the stars.

Spira's world view was predominantly Protestant but traces of the old Roman church and its concepts remained. Many people at the time were managing to accommodate contradictory systems in their minds, like the fact that Calvinism accepts predestination to hell, while the Catholic Church states no one is predestined to the eternal fires. In Spira's case, a guilty conscience ratcheted up the tension. Then came a family scandal. Mental stress fractures started to show. His delusion of despair (to non-believers anyway) presented itself to the world.

Spira gave up so much for his delusion, finally even his life, suggesting clearly that it was easier to believe himself damned than to walk the tightrope between grace and hellfire every day. James Tilly Matthews created the Air Loom to explain the turmoil he was in. Burton clung onto his horoscope. Spira submits to despair. The picture is more complicated, though. Spira's story is affecting because it shows how absolutely an individual can become locked into a delusion. The alternative reality Spira retreats into has grand theological dimensions. It removes from all sorts of complex worldly problems, but his end is a painful and sad one and demonstrates how ultimately self-destructive an alternative reality can be.

'Charles VI bedridden and his physician',
Chroniques de Jean Froissart, Book IV, 1389–1400.

CHAPTER V

The Glass Delusion of King Charles VI of France

26 February 1382, the thirteen-year-old Charles VI of France rides into Rouen, accompanied, because of his age, by his regent, the Duke of Anjou. This is his first real test as king, just two years into his reign. The city, the second largest in the kingdom, is in revolt at the reimposition of taxes by the new monarch, and he is here to put down this uprising and stop any further damage to public buildings after the destruction of the city's tax records.

The revolt is led by a journeyman draper who has assembled a mob of fellow workers. We are right in the middle of the Hundred Years War against the English Plantagenets. Rouen's population has also endured recurrent bouts of plague. The new commodities taxes are the last straw.

Charles assesses the damage. Rouen, sitting on the river Seine in Normandy, is a prosperous regional trading centre. The cobbled streets around its magnificent cathedral groan with trades premises in half-timbered buildings, all the shopfronts closed up today because of the riots. Coils of smoke and angry shouts cut the grey winter sky.

Charles and his entourage gather the names of the ringleaders and oversee the executions.

One of the smoke trails rising behind the town square is not a product of the unrest. If you were to follow it, it would lead you to a pair of locked gates at the end of an alley in the heart of the old city. Behind these gates sits a row of industrial workshops specialising in the manufacture of a new product. If Charles had cared to leave his attendants and wander up this alley, even approach the gate and take a look through the keyhole, across the courtyard, and through the partially open doors leading off it, he would have seen a series of furnaces glowing in the recesses. Pots of wood ash, too, and of sand. Four-foot-long blowpipes propped against the wall.

The insignia above the gates show that these are successful family businesses. Smartly dressed men just behind the gates guard the premises. Before the keyhole is blocked by one of these guards, the king might catch a glimpse of the disc bullions of plate glass laid out for despatch, each one carefully wrapped and stacked on a cart. Diamonds of thinner glass wait in other piles across the courtyard, to be matched with lead lattice windows in the houses of the gentry. They say you can see through it but it's strong. It would break if dropped. The artisans are still at work here, apparently not part of the uprising. Some are hiding from it. They have been targeted that day along with the other wealthy institutions.

Glass has already existed for a thousand years, and Charles is familiar with it. First there's his father's drinking vessels, and then the stained-glass windows at cathedrals like Chartres.[1] Glassmakers mastered this 'crown' glass in the 1340s and it is a particularly useful variety. It can be used to keep fresh food, for magnification and, most commercially, to protect buildings from the elements while providing an expansive view. Diamond shapes are cut out of the thinner portion of the glass disc and mounted in a lead lattice for larger, showier windows.

It is conceivable that the young Charles came across the glassworks on his excursion to Rouen. Given that they were an important centre of industry in the city, he might have been introduced that day to a foreman. He could then have made the connection between the artefacts coming out of the furnaces and the diaper glass in the windows of the palaces. This very new, alchemical work was going on, just out of sight.

The Rouen methods were a highly prized trade secret. The few families who shared it acquired a status almost equal to nobility. The clandestine nature of the operations kept the production local and meant that this particular kind of glass was not available in London for another three centuries. There had been a rival centre of glass in Murano, near Venice, from the 1290s, but the exceptionally clear glass for which Murano would become world famous was not invented until 1500. Plate glass of the kind made in Rouen was, at that time, the clearest ever seen.

This collision of place and time – the young Charles in Rouen at the same time as this art was developing – is noteworthy given what happened a few years later.

This very same French king, Charles VI, became the first publicised case of a person who believed his whole body, not just a part of it, was made of glass.

Aeneas Sylvius Piccolomini, later Pope Pius II, recorded in his chronicles that Charles refused to allow people to touch him, and wore reinforced clothing to protect himself.[2] He was reported to have wrapped himself in blankets to prevent his buttocks from breaking.

Charles VI's reign was dominated by countless battles with the English over who was to rule the kingdom of France – the largest in Western Europe at the time – as well as by internal feuding within the kingdom itself. One of Charles' recent predecessors, John the Good (John II), had been captured by the Black Prince, the heir to the English throne, at Poitier and taken back to England to be ransomed. It was precarious ruling France at this time. Charles would be defeated by Henry V at Agincourt, and while the age of chivalry was at its height, French society itself was fragile and threatening to fragment. The appeal of glass, a wondrous new material, as a metaphor for a person's sense of themselves, is not difficult to understand in this context. To be made of glass is to be precious – rare, but fragile; it is an instruction to others to admire you, but not to get too close. There is something essentially magical, spectral about glass – it's the stuff of fairy-tale glass slippers, you can see through it, but you know it's there. It's hard but technically somewhere between a solid and a liquid. Cathedral windows are thicker at the bottom than at the top, giving the impression that the material has continued to flow imperceptibly (it's actually because the glazers put the thicker edges of the glass at the bottom) and the glass has a life of its own.

The glass legs of Charles VI of France spoke eloquently of the chaotic political situation he found himself in. Glass perfectly symbolises a sense of both supreme refinement and a nail-biting sense of vulnerability and contingency. But Charles did more than champion the poetic possibilities of glass. Many people found it an appealing metaphor for the ambitions and fears in their lives. He turned himself *into* the stuff. What was influencing him? What brought about a glass *delusion*?

Various physicians and historians chronicled Charles VI's life and traumas, attempting to trace the cause of his mental instability. The religious order of St Denis claimed to be first-hand witness to many formative episodes. Jean Froissart gave his spin a few years later in a colourful illuminated prose history of the Hundred Years War, complete with lively pictures, and the cleric and historian Jean Juvénal des Ursins produced a more sober history.

According to the order of St Denis in April 1392 the twenty-four-year-old Charles VI was in Amiens when he suffered from an extremely high temperature that lasted for weeks, probably as a result of contracting typhoid.

His mental state became increasingly volatile: when out on an expedition on 5 August 1392 he apparently saw a man dressed in rags in the forest in Le Mans. Froissart says the apparition pursued the king for nearly half an hour, resisting the many armed men who were with him, even surviving several blows of the king's sword, before disappearing as inexplicably as he arrived. Charles apparently became frenzied, paranoid, lashing out even at his own soldiers, crying out: 'they want to hand me over to my enemies.' He spurred his horse and rushed from the scene.

The supposed eyewitness in the order of St Denis remembered it a bit differently. It wasn't an apparition, but a 'a wretch'.[3] Froissart described 'a man, head and feet bare, dressed in a beautiful burel dimension',[4] and Juvénal embellished the story with an account of the king killing four of his knights in his frenzy as he left the scene.[5]

Charles apparently didn't remember any of it, and, according to Froissart, suffered another high fever.

On 23 January 1393 Charles was involved in a horrific accident at a masked ball. A celebration had been staged for a wed-
ding and a group of men dressed as hairy satyrs to perform an
entertainment were set on fire by a stray naked flame. Charles
was among these satyrs as the fire raged. He was whisked away
from the carnage of what became known as '*le Bal des Ardents*',
'the Ball of the Burning Men'.

Another episode heralded the start of his glass delusion. A
member of the order of St Denis remembered the occasion in
August 1395 when Charles, now twenty-six, failed to recognise
his own family. We might think of it now as a 'disorder of recognition', to use Joseph Capgras's expression. *Et qui n'est chaque fois
ni tout a fait la même, ni tout a fait une autre.* The fragment of poetry
that captures 'Madame M's confusion at her daughter's doubles
might be fitting for our king, too. Charles became agitated at his
own distorted perception, manic, and it was out of this state that
his glass delusion first emerged:

> He often ran two and fro in his palace, until his strength was
> completely exhausted… The king also claimed that he was
> made of glass, clad himself with iron splints, feared breaking
> when falling, etc. And he supposed himself sometimes to be
> of glass, not to be touched, he was suffering. He inserted the
> clothes on a rod of iron…[6]

Charles's condition is depicted in an illustration to Froissart's
fourteenth-century chronicles. He is in bed, attended by courtiers,
one of whom presents him with some sort of elixir. It is contained,
appropriately enough, in a precious clear-glass decanter. He's

been handed down to us like this, a caricature of a 'mad king', with cushions tied to his backside. This is Charles as a symbol of the lottery of primogeniture, in a line of 'mad kings', like Lear raging in the storm. If you're going to have a hereditary, divine right monarchy, the image says, this is the flipside. Traces of Charles the man are harder to find, never mind a nuanced picture of his formative experiences and influences. We are left with a very few (presumably apocryphal) anecdotes to patch together.

The details of Charles's life get lost in the phenomenon that glass delusion became, and that he came to represent. It's a phenomenon that lasted for centuries, and spread way beyond the French court. Cases grew over the sixteenth-century Europe and continued to flourish well into the seventeenth.

The people recording the story as it unrolls refer back to Charles again and again. He's a puzzling case and still the most renowned sufferer of glass delusion more than two hundred years later when Burton began compiling his encyclopedia of melancholy. At this point cases were concentrated among the wealthy and educated classes of men (although, of course, physicians rarely saw other categories of people). This led contemporary scholars to associate it with the better-known disorder of 'scholar's melancholy'. Sufferers were observed to be normal in every respect, bar the belief that they had turned to glass. As a result, they could function relatively well, albeit remaining anxious in case anyone came too close and risked shattering their fragile limbs.

For the aristocracy this was a 'delusion of grandeur' in reverse. 'Madame M' believed she was related to King Louis. A bona fide princess, Alexandra Amalie of Bavaria, believed she had swallowed a glass piano and insisted, for obvious reasons, that people should not get too close. It caught the private imagination of non-royals, too. In 1583 Tomaso Garzoni reports a Glass Man travelling to Murano. He plans to fling himself into a kiln and be transformed into a goblet.[7] Another case features a scholar who believed that the surface of the world was made of glass, beneath which lurked a tangle of serpents. He did not dare leave his bed, fearing he would smash the glass and fall in among the snakes. Painters of 'Vanitas' put hourglasses into still lives to remind viewers of the fragility and transience of life.

Distressing material changes in the body often feature in the content of delusions but 'glass delusion' was different. Robert Burton's includes many 'hypochondriacal' delusions of the body in *The Anatomy of Melancholy*, like the baker in Renaissance Italy who thought he was made of butter and wouldn't go near his bread ovens, or man who 'thinks he is a nightingale, and therefore sings all the night long', or the people who are 'all cork, as light as feathers; others as heavy as lead; some are afraid their heads will fall off their shoulders'.

Glass delusion is different. In Burton's schema, glass delusion sits with a sketchier subset of examples with the people who 'Fear of devils, death, that they shall be so sick, of some such or such disease, ready to tremble at every object, they shall die themselves forthwith, or that some of their dear friends or allies are certainly dead; imminent danger, loss, disgrace still torment others'. Here are those who 'are all glass, and therefore will suffer no man to come near them'. They are with the people who think they are already dead, who negate themselves and become passive, who stop eating. The historic enigma of Charles VI hides somewhere

here, along with possible explanations for him in the broader story of glass and how others reacted to it.

The Franciscan friar and scientist Roger Bacon had described the marvellous properties of the magnifying glass back in the thirteenth century. Glass in its new forms maintained a connection to developing technology, from spectacles to astrological instruments.

By the time glass delusion was at its peak, the science of optics was really taking off. This was the period marked by comets, solar eclipses and conjunctions of the major planets that Burton gazed up at, and studies required specialist optical instruments. The light-accommodating properties of glass also became associated with clairvoyance.

The question of prophecy was a sensitive one, and 'magical' lenses were officially frowned upon. James Howell, historiographer to King Charles II and glassmaker by trade, writes in the seventeenth century about the properties of glass. As he stares into a glass kiln, Howell reflects on the alchemical nature of what's going on in front of him, and sees a metaphor for the earth's place in the grand scheme of things, destined to be consumed and purified by the holy fires.

> ... it being a rare kind of knowledge and chemistry to transmute dust and sand (for they are the only main ingredients) to such a diaphanous pellucid dainty body as you see a crystal glass is, which hath this property above gold or silver, or any other mineral, to admit no poison... My thoughts were raised to a higher speculation: that if this small furnace fire hath virtue to convert such a small lump of dark dust and sand into such a precious and clear body as crystal, surely that grand universal

fire which shall happen at the day of judgment, may by its violent ardour vitrify and turn to one lump of crystal the whole body of the earth; nor am I the first that fell upon this conceit.[8]

Francis Spira's death bed of despair was only fifty miles from Murano, by now the centre of the glass world. He would have known all about glassmaking. Did he see an allegory of damnation and redemption in the process, too? Spira died only five years after Copernicus shocked Europe with the theory that the earth went round the sun and he didn't escape difficult new questions about man's place in the universe. He missed the revolution in telescopes that Burton lived through. Burton's interest in clairvoyance would have given him a natural interest in glass lenses, luminous portals to the unknown.

The historian Gill Speak argues that the phenomenon of glass delusion had become so well known and widespread by the seventeenth century that it was possible to talk about Glass Men as a discrete set within the population.[9] In line with Howell's vision, Speak suggests these Glass Men saw themselves as the product of a 'purifying fire'. Glass delusion was a symptom of the prevailing religious anxiety about death and the afterlife. A contagion of ideas, and of beliefs, could spread faster than at any time in human history, thanks to the speed with which books could now be copied.

At the same time, glass delusion had become a fashionable literary motif. There were a multitude of references in fiction, the highest profile example being Cervantes's short story 'The Glass Graduate' of 1613. The hero of the tale, Rodaja, is a brilliant student at Salamanca University who is poisoned by a quince, a potion given to him by a worldly lady who has sought his love in vain. She intends the potion to serve as an aphrodisiac, but instead it triggers an illness which lasts a year. It is an illness that develops into a delusion that he has turned into glass. 'The Glass Graduate' was republished in Latin, and before the end of the seventeenth century it had been translated into French (1615), Italian (1626), German (1652) and English (1694) and made its way across Europe.[10]

The German humanist Kaspar von Barth is singled out most often as the model for Rodaja in the Cervantes story. Von Barth reportedly shunned company for ten years, believing that he was made of glass. Cervantes may also have known about an unnamed royal mentioned by the physician to Philip II of Spain who covered himself in straw to avoid smashing (many sufferers went for wool or straw, for self-evident reasons).[11] The court physician apparently devised a 'ruse', a little trick or white lie, setting fire to the straw. The royal started banging on the door, begging to be released. The physician let him out and then asked him why, if he was made of glass, he hadn't broken into pieces with all the knocking? The man was instantly cured, or so the physician said.

There are other accounts in which the delusion is simply observed. André du Laurens, a friend of Philip II's physician (who also mentioned the same unnamed royal) described a melancholic in his *Discourses of Melancholike Diseases* of 1599 who 'prayed all that came to see him not to come neare unto him, least they should dash him in pieces'. There was also a 'great Lord, which thought himselfe to bee glasse, and had not his imagination troubled, otherwise than in this onely thing: he used commonly to sit, and tooke great delight that his friends should come and see him, but so as that he would desire them, that they would not come neere unto him… Another imagined that his feete were made of glasse, and durst not walke least he should have broke them.'[12]

This lord has one fixed, false idea about himself, but is otherwise behaving within the generally accepted rules. His glass delusion shows its workings here as an effective strategy for social anxiety. Charles VI of France is part of a sizeable group whose glass delusions challenge the idea that royalty and the aristocracy are above that sort of thing simply by accident of their birth.

The phenomenon moves all over the place to other social groups and many cases are serious afflictions. The Amsterdam-based poet and philosopher Caspar Barlaeus, a near contemporary of Robert Burton, talked in his correspondence about his chronic melancholy and the attendant belief that he was made of glass was debilitating. There are tales of people living with glass bones, heads, arms and hearts.[13] Sitting down is still something to be afraid of for some people in France centuries after Charles VI. According to a report by Gédéon Tallement des Réaux in 1657, Nicole du Plessi, a relative of Cardinal Richelieu, was terrified that her rear end would shatter.[14] The accounts are strikingly consistent. People made of glass all take the same precautions and demand the same care, respect and physical space.

Walkington's *Opticke Glasse*, published in London in 1607, features a melancholic Venetian who, just like Charles, is so convinced that his shoulders and buttocks are made of glass that he is afraid to sit down and risk breaking his 'crackling hinderparts'. Also like Charles, and Cervantes's Rodajo, he's experiencing active persecution. He doesn't venture out in case a passing glazer catches him and melts him down for use in a latticed window.

There were many who refused to take the delusion entirely seriously. Many comedies were written on the subject.

In Thomas Tomkis's play *Lingua, or The Combat of the Tongue, and the Five Senses, for Superiority* the trickster Tactus is feigning an attack of melancholy and describes the moment he is transformed into glass:

> No sooner had I parted out of doors,
> But I held up my hands before my face,
> To shield mine eyes from th' light's piercing beams;
> When I protest I saw the sun as clear
> Through these my palms, as through a perspective.
> No marvel; for when I beheld my fingers,
> I saw my fingers were transform'd to glass;
> Opening my breast, my breast was like a window,
> Through which I plainly did perceive my heart
> In whose two concaves I discern'd my thoughts
> Confuse'dly lodged in great multitudes.[15]

Tomkis lampoons the delusion as an affectation and a ploy used by charlatans, just as Shakespeare exposed the fraud of the astrologers. Another sense, Olfactus, joins the debate:

> See the strange working of dull melancholy!
> Whose drossy thoughts, drying the feeble brain,
> Corrupts the sense, deludes the intellect,
> And in the soul's fair table falsely graves.
> Whole squadrons of fantastical chimeras
> And thousand vain imaginations,

> Making some think their heads as big as horses
> Some that th' are dead, some that th' are turn'd to wolves,
> As now it makes him think himself all glass.

Tactus protests:

> Olfactus, if thou love'st me, get thee gone;
> I am an urinal, I dare not stir
> For fear of cracking in the bottom.

Tactus is a glass object – to piss in. He's a joke. But there's also a more ominous note. Melancholy here is an insidious force that completely takes a person over, manipulates them and wrecks their judgement so that they think they are wolves, or even that they are dead.

In 1641, the Dutch poet and friend of the philosopher René Descartes, Constantijn Huygens, wrote a popular poem, 'Costly Folly', whose subject fears everything that moves:

> What's wrong? Well everywhere he's touched is made
>   of glass, you see;
> The chairs will be the death of him, he trembles at the bed,
> Fearful the one will break his bum, the other smash his head;
> Now it's a kiss appals him, now a flicked finger shocks,
> Just as a ship that's gone off course sails fearful of the rocks.[16]

This is a pathetic character, too cowardly even for a kiss. But he's also in mortal danger, like a vessel adrift in the sea. Glass delusions, as with Charles VI, are never straightforwardly funny – they are too disconcerting. Incidentally, Huygens's son Christian links several of the subjects in this book. He was a brilliant mathematician, astronomer and physicist whose work featured in several of the delusions. In 1659 Christian discovered the true shape of the rings of Saturn, thanks to his work grinding and polishing lenses for telescopes. He also contributed to the science of dynamics – the action of forces on the body – as well as founding the wave theory of light. He even invented the pendulum, which our headless clockmaker (see Chapter VII) will try to exploit for energy in his perpetual motion machine. He also suffered with melancholy.

Huygens's friend René Descartes, who famously divided the body and mind, is in no mood to try and understand how these 'fashionable' delusions function. In the same year as Huygens's poem his *Meditations on First Philosophy* expresses shame by association with fellow men and women who make absurd claims: 'But how could I deny that I possess these hands and this body, and withal escape being classed with others in a state of insanity, whose brains are so disordered and clouded by dark bilious vapours as to cause them perniciously to assert that they are monarch when they are in the greatest poverty...or that their head is made of clay, their body of glass.'[17] This sounds like a new spin on the old imbalance of humours, even demonic forces at work.

The delusion is endlessly satirised, but as with Charles VI, it made people uneasy and couldn't be dismissed. Glass is a muse for all times, but why did Charles turn himself into it? Plenty of people with an interest in the subject have picked over the biographical scraps trying to further that enquiry.

Someone asked the question in the mid-1920s. Dr Ernest Dupré, a psychiatrist at the Faculty of Medicine in Paris, was working on delusions. Dupré was a peer of Joseph Capgras, both men engaged in a subject that was experiencing something of a renaissance. While Capgras pondered 'Madame M', her murdered husband and the substitute doubles, Dupré fell upon the case of the 'Mad King Charles', thinking it might be a new way into the topic.

In 1925 Dupré wrote an influential book on the pathology of the imagination.[18] In it he digests the original chroniclers of Charles VI and writes up a diagnosis of the king's mental state, putting a twentieth-century lens on the story. The result naturally says as much about France after the First World War as it does that of the fourteenth century. The figure of 'Mad King Charles' gets more definition through the eyes of those people who were so stuck into the subject of delusions, pulling together the ideas and terminology from psychiatry and psychoanalysis of the previous two hundred years since Pinel. Dupré has gone back over the sources looking for telling additional details and gives his summation:

> During the attacks, we recognise in the picture two orders of symptoms… In the foreground, signs of motor and psychic arousal appear: broken objects, obscene gestures, vociferations, etc. but at other times, and sometimes, it seems, at the same time, the king manifests a despondency, a state of inertia and torpor in which, according to the text of the Religious at St. Denis, he refuses 'To change shirts and sheets, take baths, shave beard, and finally eat and sleep at set times. He remained silent for long hours'. 'We never came to see him,' said Juvénal 'and he looked at people hard and said nothing.'

Like Burton, his mental imbalance goes back a long way. We see a man who swings from being violently agitated, to depressed and withdrawn to the point of catatonia. Here again is the torpor, the retreat from life and agency demonstrated by 'Madame M', Francis Spira and Cervantes's Rodaja with his 'four-hour' faint. Dupré gives his take on the king's feelings about the deaths at Le Mans:

> At other times he suffers, laments, and searches with anguish for what may be the cause of such torments. Fearing the wrath of heaven, he sent donations to the chapters, especially to Saint-Julien du Mans 'because of the murders he committed' in the forest. According to Juvénal, he introduces and preserves in his flesh a piece of iron which produces an infected ulcer.

Like Spira, he fears God's judgement and the burning fires which will consume him. Guilt tortures him to such an extent that he stabs himself, and nurtures the resulting infection as a punishment. More than a hundred years before Francis Spira, he does something that is absolutely forbidden, a sin. He despairs.

> Feeling a fear, moreover quite natural at a time and in an environment fertile in criminal poisonings, evoking family memories, he wonders if he is not poisoned, and far from reacting with anger and recriminations, he confines himself, in despair, to begging to be finished. 'If he is here,' he said, 'the one who makes me suffer, I implore him, in the name of Our Lord, not to torment me any more, to make sure that I no longer languish and die.'

A childhood growing up in the French royal family has clearly had an impact. Charles is also paranoid. Or is he bang on the money with his concerns? He would inevitably have watched all manner of skulduggery at work in the royal household, as wars and successions were decided. There wasn't just capture by the English to fear. Charles has learned the hard way even to be suspicious of what you were given to eat at home in case it was your last meal. Dupré presents the pathology:

> To these manifestations of inhibition and depression, it is advisable to compare a set of morbid conceptions that Charles VI, during these crises, frequently expresses: they are delusional ideas of negation and transformation, concerning his personality: he doesn't have a throne, no coat of arms, no wife, no children. His name is George, he's made of glass, etc. etc.[19]

Like Spira, Charles retreats. He refuses his identity, doesn't recognise his wife or children. Turning himself into glass is just one part of the project of withdrawing from all the ongoing troubles. It's process of stepping away from himself which sees his identity unravel, throwing off all the trappings that make him who he is, in public at least: a husband, a king, a son, a father, even a 'Charles'. He leaves behind a glass replica of his body.

In search of the cause of the delusion Dupré now looks past Charles's early environment, to what he has inherited through the bloodline. He gives a new name to what Burton would have called melancholia:

a heavy morbid heredity is first of all obvious: there exists, especially in the maternal line, many psychopathic antecedents of a mainly depressive nature.

Dupré points to a rumoured history of depression in the female line. Charles's mother, Joanna of Bourbon, married Charles V when they were both twelve. There are second-hand reports that she experienced episodes of mental instability and had some kind of breakdown when her son Louis, Charles's younger brother, was born in 1378, when Charles would have been three. Significantly, though, her husband always referred to her as a capable and trusted regent. She died following childbirth, when Charles was nine years old.

> In his youth, the King, of robust physical constitution, but of probably mediocre intelligence, debauched, lavish, and always in the grip of sterile agitation, presented himself above all as an imbalance of character and conduct. On this constitutionally insecure ground, at the age of twenty-four, a febrile affection arises (probably typhoid fever), which is accompanied by psychic and convulsive disorders... During the convalescence, which is slow and difficult, the King shows himself bizarre, fantastic, agitated and he madly undertakes this expedition to Brittany, which was to be interrupted so miserably, almost at its beginning, by the furious crisis of the forest of Le Mans.

Dupré's portrait of Charles is of a frustrated youth, not too bright but spoiled, his personality out of step with his actions. Dupré also gives a nod to typhoid, specifically the acute fever that went with it as a possible contributory factor to Charles's mental imbalance. Charles was on 'constitutionally insecure ground', as Dupré put it, but something had to have triggered the delusion. Capgras had considered the same disease significant in 'Madame M's case, and de Clérambault noted episodes of diphtheria and jaundice with interest in 'Léa-Anna B's history (see Chapter X). It's after this protracted fever that the rudderless young Charles starts to exhibit 'bizarre' and 'fantastic' behaviour.

In the nineteenth century, the Anglican priest and scholar Sabine Baring-Gould wrote an eccentric but scholarly history of the phenomenon of werewolves and the associated mythology and folklore and reports a link to typhoid fever. In it he set out the 'natural causes' of the delusion of 'lycanthropy', where a person believes that they have taken the body of a wolf. He connects typhoid to delusional states asserting that a victim of typhoid fever 'may regard his members as composed of [such] foreign and often fragile materials as glass'. Rheumatism and gout can apparently produce similar hallucinations, he says.[20] Today, it is common for patients in intensive care to experience delusional hallucinations brought about by a combination of physical trauma, drugs taken while in hospital and fever which produces a belief that the body has been transformed into something inhuman. Paranoia is also common. A disease like typhoid could well have contributed to Charles's instability and predisposed him to delusion.

Charles reigned for forty-two years, and continued to experience episodes of psychosis and delusions, that were not limited to glass, earning him that name of 'Charles the Mad'. He did have long, intermittent periods of mental stability, however, and even managed to retain some power after defeat in the war by marrying his daughter to the victor, Henry V of England. Over the years he sought remedies for his illness, in the form of exorcism, and trepanning: the drilling of small holes in his skull. But Charles VI died in Paris 1422 at the age of fifty-three, without finding a cure for his mental turmoil.

It's clear there were many factors which could have led to Charles's mental crisis, genetic predisposition, traumatic events, fever among them. The picture is incomplete. The image of Charles with his backside wrapped in blankets is the icon of the golden age of Glass Men which followed. Cases then disappear almost completely from the records.

Modern research has turned up a few recurrences. The French translator Raymond Foulché-Delbosc reports a Glass Man in a Paris asylum in 1892.[21] Then a lecture of 1883 from the archives in an Edinburgh mental hospital yields a case which is significant because it suggests that glass delusion was something experienced by women, too.

It was the American philosopher and psychologist William James who turned up the case. He came across Dr Clouston's 'Clinical Lectures on Mental Disease' at the Royal Edinburgh Asylum, the institution Clouston ran, and wrote up the lectures up in *The Principles of Psychology* in 1890. In the third lecture of the series, Clouston lists the symptoms of 'about a hundred female melancholy patients'. James takes a moment to point out that delusions must be distinguished from illusions and hallucinations, writing: 'A delusion is a false opinion about a matter of fact, which need not necessarily involve, although it often does involve, false perceptions of sensible things.' Then he's back to the job in hand, listing the wide variety of delusions that the women presented to Dr Clouston with, which included: 'general persecution', 'having no stomach', 'being followed by the police', 'the soul being lost', 'that the head is severed from the body', 'unseen agencies working', 'having neither stomach nor brains', 'her children being killed', 'being a fish', 'being dead' and 'the legs being made of glass'.[22]

Glass delusion had not disappeared at all. It had simply gone underground.

A few more cases of glass delusion pop up in modern medical literature. In 1944, the French psychologist C. Pottier cited the case of a woman who thought that her chest and right shoulder were made of glass, and that she had heard the bones of her face cracking.[23]

Yet another case is found in an aside within an article profiling a Turkish-Greek military engagement during the First World War. An anonymous byline in *Time* magazine of 3 February 1958 referenced Atatürk's successful offensive. The report went on to explain the Turkish victory not in terms of their superiority but in relation to the Greek general's belief 'that his legs were made of glass and would break if he stood up moved about too freely'.[24]

There's a mention of glass delusion in passing from the essayist Erich Heller in 1958, who drops in by the by that he 'once heard a strange case of mental affliction. A distinguished writer at breakfast one morning refused to sit down, protesting that an important part of his body was made of glass and was bound to break on impact with the chair. On another day, though, his nurse found him comfortably seated in a chair. "You're cured!" she cried. "I knew it would happen! Such an intelligent man…" "Cured?" he replied. "Of what? Perhaps of a slight misapprehension. Why did nobody ever tell me that it is unbreakable glass?"' Heller then adds a bit of gossip about what happened in the aftermath of this exchange: 'He was rightly released as being perfectly adjusted to life, but unfortunately among those who knew him there lingered a slight doubt which even affected their reactions to his distinguished books.'[25]

The history of glass delusion is fragmented; a trail of anecdotes. Psychiatrist Andy Lameijn's research in the 1990s led him to an encounter with a patient who was experiencing a glass delusion first-hand. Lameijn's account offers rare access to a personal conversation about the delusion and compelling detail about what it's like to experience.

Lameijn was director of the Endegeest Psychiatric Hospital in Leiden in the Netherlands when a fellow psychiatrist approached him. Years earlier, Lameijn had written and lectured on the subject of historic 'glass delusion' as a medical curio and his colleague thought he might be interested. Lameijn was aware of the story of the Mad King of France and the phenomenon of the Glass Men, but the case in point was one of glass delusion from the 1930s in the archives of another Dutch hospital. A woman had been admitted to a psychiatric hospital believing that her legs and back were made of glass. Such was her fear of personal contact, the notes showed, that the nurses couldn't get near her to change her clothes or help her. She apparently recovered following treatment. A few months later, another doctor brought Lameijn a case from a different hospital from 1964, and shortly after that a young man turned up at the University Clinic in Leiden, claiming to be made of glass. 'I really dropped everything,' Lameijn recalls, 'I didn't want to miss this.' He was going to have the opportunity to speak to someone, apparently the first in decades, who was experiencing glass delusion.

Lameijn talked for several hours to the man. He asked him what this feeling of being made of glass meant to him. Not wanting to distort the conversation, he didn't bring up ideas of fragility or transparency. Initially reticent, the patient then began to open up. He pointed to the window in the consulting room and asked Lameijn what he could see. Lameijn replied that he could see a street, some cars, more buildings, people walking past, and waited.

The patient said: 'Ah! You've missed the glass in the window. You didn't see it. But it is there.' He leaned forward, and said: 'That's me. I'm there, and I'm not there. Like the glass in the window.'

The conversation continued, and the patient expanded a little on his feelings about being made of glass, claiming that he was able to turn this feeling of being 'there' and 'not there' on and off at will; as if he had a switch in his own mind he could 'disappear' and 'reappear'.

It transpired that the patient had recently been in a bad accident, and Lameijn began to formulate a theory as to why in this day and age a person might present with glass delusion. Glass was not a new invention. He concluded that the man in question was using his delusion as a sort of distance regulator. Following the accident, his family had become overprotective, and the glass delusion was his attempt to regain privacy and hide from his overbearing family.

Again, identity is unstable, faces familiar and unfamiliar, like 'Madame M's doubles. Here's another person withdrawing from the world via their delusion, just as Spira, and, for different reasons, Charles VI did. There are echoes of 'Madame X', a woman whose bedside we'll visit in Chapter IX who is convinced she has been dead for some time.

Why is glass delusion so persistent? What has kept people returning to it? There is good reason why someone experiencing mental disturbance in early modern Europe might adopt glass for a delusion. Here was a hard yet transparent material created by heating rock: beautiful, magical, breakable. The relative newness of clear glass as a material in seventeenth-century Europe explains how it caught the eye.

New technology features again and again in delusions all the way through. It comes to dominate the delusions of the 1950s and 1960s. As news of ingenious Cold War spying devices broke round the world, US doctors saw a spike in cases of people asking for help because the CIA had inserted a chip into one of their teeth and were using it to download their thoughts. But glass was no longer new by the time Andy Lameijn met his patient in Leiden.

More awkward questions remain. Why is it glass that appears in delusions related to technology, and not the steam train? The guillotine and not the printing press? Certain new technologies are more obviously disconcerting to encounter for the first time. We can see how the appearance of glass might disturb someone who views it as alchemy. The guillotine would chill anyone's blood at first viewing.

But the persistent tropes like glass offer something that outlives their newness. Glass has many qualities, referencing religion and mysticism as well as technology. A person can easily map their own complexity onto glass, synthesising themselves with it.

Glass can package up complicated and conflicted personalities in a single, tangible image for all the world to see. It is both precious and breakable just as the guillotine is a symbol of human ingenuity, and also of the knife-edge of life and death. Glass can take on all the many influences that come together to create mental confusion and make sense of them, even make something magical out of them.

The psychological dimensions to glass delusion are complex. The Glass Men frequently believed that they were being persecuted, and show a hypervigilance that looks a lot like paranoia. Rodaja in Cervantes's tale fears persecution in his extra fragile state and sleeps in haylofts, wearing loose clothing and avoiding the locals who throw stones at him. 'He walked always in the middle of the street lest some tile fall upon him…and during the winter, buried himself in straw. When his friends embrace him to prove he won't break, he falls into a four-hour faint.' There are echoes here of James Tilly Matthews running from the Air Loom Gang and of 'Madame M' evading the doubles who want to poison her. Glass is a natural communicator of paranoia. A person made of glass can be seen into, and 'read'. But with a glass delusion there is a contrary sense that the glass person is inviting the observer in and that it's not a violation so much as a request. Transforming into glass doesn't solve everything by itself, of course. There are demands attached to a glass delusion, most urgently how the person with the delusion must be treated – with careful attention, care, respect. It's never clear if the demands will be met. The signs are that in most cases they won't be.

Glass is still resonant. Anxieties about fragility, transparency and personal space are more pertinent now to many people's experience of living than they've ever been. It's an elegant response to the society we have to navigate; a society that is increasingly crowded, in which modern technological advances isolate us but also offer apparently boundary-less communication. Glass delusion expresses social anxiety, which many of us experience to a lesser extent. It allows us to demand protection against drops and shocks, and transforms a lost soul like Charles VI into something worth treasuring and nurturing.

Portrait of Margaret Nicholson when old,
John Thomas Smith, 1800–1828.

CHAPTER VI

Margaret Nicholson, Descended from Boudicca and Rightful Queen of England

The morning of Wednesday 2 August 1786, St James's Palace, London. A short, respectably dressed, middle-aged woman loiters outside the royal residence, making casual conversation with two other women she's just met. According to an early press report, she is in her best clothes, wearing a 'flowered linen or muslin gown, black gauze bonnet, black silk cloak' and a 'morning wire cap with blue ribbons'.

After an hour or so, the king, George III, approaches in his coach. The woman holds up a piece of paper and the king stops at the garden door, opposite the Duke of Marlborough's Wall, alights from the coach and approaches her, as he customarily does when subjects come to him asking for help, although the woman is allowed only within arm's length. A special edition of the *London Gazette* reports that she is 'rather a little woman, of a swarthy complexion, a native of Durham, her father a barber'. Another paper wades in, guessing she is about thirty-six and calling her a 'native of Yorkshire' with 'the appearance of a foreigner'. Only the last of these facts will prove to be true. She offers the piece of paper to the king – it's a petition – and as he stoops to receive it he feels a thrust at his belly, passing between his coat and waistcoat. He draws back: 'What does this woman mean!' The yeoman jumps in, and a knife is released from the woman's other hand. The only damage done is to a button on his waistcoat. The 'petition' is blank. The king apparently takes stock, then declares: 'I am not hurt! – take care of the woman – she is mad – do not hurt her.'

The woman is apprehended but giving nothing away. She is 'taken to the Queen's antechamber where she remained from twelve until near five, during all which time, though spoken to by several of the nobility, she did not condescend once to open her lips, but appeared totally unmoved'.

According to the *London Gazette* that week of 1786, the event had quite an effect on the king: 'His Majesty then went forward into the palace; and, when he had recovered his surprise, appeared to be greatly affected, expressing in a kind of faultering voice, that, "surely! He had not deserved such treatment from any of his subjects."' Much will be made of the mercy of the king in his response to Margaret when compared to the French, who have recently dealt with a similar attempt on Louis XVI's life but labelled it regicide, rather than insanity, and were brutal in their response.

The woman is interrogated and offers her real name: Margaret Nicholson. She says that she will give her reasons when she is 'brought before proper persons'.

The story is turned around quickly and spun breathlessly for several papers, but the *London Gazette* of August 1786 trumpets its more or less instant response, getting the truth out there to stop the 'mischievous effect' of the 'thousand fictions' that were already circulating.[1] The paper breaks the suspense with a spoiler – the king is alive and well – and is forensic with 'Particulars of Margaret Nicholson's Attempt to assassinate His Majesty'. It makes a fuss of the trivial details – the knife was not hidden inside the petition, as widely reported, but under her cloak. But then there's a crucial point about the knife itself. The paper says the 'instrument she used was an old ivory handled dessert knife, worn very thin towards the point; so thin, that a person pressing the point against his hand, it bent almost double without penetrating the skin'. The knife, in other words, was not a lethal weapon. By dialling down the sensation this source becomes the closest to a credible account. Nicholson hadn't intended to kill the king.

The interrogation is witnessed by the prime minister himself, William Pitt (this is ten years before Pitt bears witness to James Tilly Matthews's allegations of a treasonous element in his government, and two years before George III's own mentally instability throws his administration into crisis). Also in the room are the Earl of Salisbury, Francis William Drake of the Royal Navy, a descendant of the brother of the Elizabethan hero, and assorted magistrates. Nicholson begins an account of herself. She came to London at the age of twelve and lived in several 'creditable services'.[2] She measures her words carefully, but it's not long before the issue of her rightful inheritance comes up. Her composure breaks: 'she went on rambling, that the Crown was her's [sic] – she wanted nothing but her right – that she had great property – that if she had not her right, England would be drowned in blood for a thousand generations. Being further asked where she now lived, she answered rationally, "At Mr. Fisk's, stationer, the corner of Marybone [sic] Wigmore-Street".' After an apocalyptic vision she's back down to earth with the perfectly ordinary details of her residential address.

Mr Fisk is duly sent for and interrogated. This is the stationer's point of view via the *London Gazette*: 'she had lodged with him about three years; that he had not observed any striking marks of insanity about her – she was certainly very odd at times – frequently talking to herself – that she lived by taking in plain work &c. Others who knew her said, she was very industrious, and they never suspected her of insanity. Mr Monro being sent for, said, it was impossible to discover with certainty immediately whether she was insane or not.' Mr Monro is in overall charge at Bedlam and the latest in a dynasty of Monros running the hospital. John Monro will appoint John Haslam as apothecary in 1795 and both men go down together after the public enquiry into abuses at the hospital in 1816 thanks, in part, to the testimony of James Tilly Matthews.

Fisk the stationer and the other lodgers all say they found Nicholson 'quiet and respectable'. To her landlord she 'always appeared a harmless character and…although she has frequently seemed in a state of absence, he never received greater proofs of his insanity in her than frequently moving her lips as if talking, and appearing agitated although in no conversation with any person'.[3]

After the flurry of newspapers came the books and pamphlets, widely circulated, their content taken largely from the newspaper reports. These drove a feeding frenzy of gossipy titbits about a woman who said she was the rightful monarch. At least five 'chap-books' came out that August, the most substantial of these street pamphlets being *Authentic Memoirs of the Life of Margaret Nicholson*, and *The Plot Investigated*.[4] Her landlord and character witness Mr Fisk turned out one of the very first. He published his version astonishingly quickly, just days after the event, on 15 August 1786. Being a stationer, he was well positioned to do this, and sizeable profits followed. The two other accounts of Nicholson vied with Fisk's to corner the market for news and background, but Fisk had the edge because of his close acquaintanceship. On the cover, just after his name, Fisk had proudly written: 'with whom she lodged'. Nicholson quickly becomes a commodity.

All iterations of the story agree on something. Her claims were grandiose. She admits that she was a servant until recently, but she wants to stress that this was with an aristocratic household. She remains adamant that she is rightfully the Queen of England and that George III has usurped her. She also pleads mitigation. She did write to her usurper but the letters were sadly ignored. Like James Tilly Matthews, she has been forced to take direct action. She insists her intention was only to frighten the king, and she thought the shock of the knife would do the trick.

Ever more regional papers, books and pamphlets join the fray, rehashing and embellishing the story. A feedback loop gets going between the three chapbooks and the newspapers, each quoting the other, and referencing the notes from 'The Examination of Margaret Nicholson by the Privy Council', until it's unclear who said what in the first place.[5] The *Hereford Journal* backs up some of Nicholson's claims. She apparently 'lived formerly in Lord Coventry's family as an attendant to some of his Lordship's daughters'. There's been a reduction in circumstance to a much less comfortable day-to-day existence and 'since that time she has existed as a sempstress, in the millinery and mantua branches'.

Her claims to an inheritance have surfaced before: the Privy Council learns that six years earlier she lived with Miss Price of Argyle Buildings, whose service she 'quitted on a pretence' of a 'a capital fortune'. She also lived with a hatter – the *Hereford Journal* of 10 August names him as Mr Watson of New Broad Street – whom she repeatedly told she 'had a large claim upon government'. Mr Watson is called in and so is a woman called Ann Southby who, the *London Gazette* says, '"lodged in the next chamber" to Margaret'. Margaret does not appear 'in the least embarrassed' in front of the Privy Council and specifies again and again exactly what she's owed. She talks of this 'claim on government – "law suit" – "just cause" and suchlike sentences'.

Everyone who knew her well seemed to agree that she was a hardworking, modest woman who had taken in work as a seamstress, to make ends meet since leaving service. 'Her brother is a respectable character according to the papers and keeps the Three-horseshoes public house, the corner of Milford Lane, in the Strand.'

Margaret Nicholson's proud attitude suggested a woman with a scrupulously well-kept home. The reality was somewhat different.

Westminster magistrates went to her apartment to see what they could find but '[N]othing more could be traced than scraps of papers, in which the names of Lord Mansfield and other persons of consequence appeared, with some disjointed writing, mentioning effects, and what she denominated "classics", a term she did not seem to understand'.[6] Beneath the snobbery are signs of a disordered life and fragments of an education. 'When the magistrates came to search her lodgings, they found nothing but three letters…and in her pockets three half-pence, and a silver six-pence, all the money she had; and as to cloaths, those on her back were her whole stock, and, except her cloak and bonnet, were very indifferent. Lord Sydney ordered her cloaths, and all other necessities, of which she was in great need.' There'd been no safety net to catch her when she left her position. She was destitute.

What on earth to do with Margaret Nicholson? The Archbishop of Canterbury, the Lord Chancellor, the Attorney and Solicitor General, and Doctors Monro Sr and Jr from the Bethlem Hospital join the Privy Council to chew over the question.

The Council decide to send her to lodge with a Mr Coates, one of the King's Messengers. Mr Justice Addington visits her there and she repeats her complaint. The king has no right to the crown. It is hers.

9 August 1790, Coates takes her by hackney coach to Bedlam, in Moorfield, north of the City of London, accompanied by a nurse and this King's Messenger. They tell her they are going to a 'party of pleasure' and she readily agrees to the excursion.[7] She 'was in very good spirits', says the *London Gazette* 'and talked very rationally the whole of the way… Upon her entrance into Bedlam, she was asked, if she then knew where she was. She answered, "Perfectly well." The steward of the hospital behaved with much kindness to her, and invited her and the company to dine with him, which they did, and during the whole time she appeared perfectly collected, except when the name of the King was mentioned', at which point she said that 'she expected him to visit her'.

She is admitted without a fight and taken to the west wing, where the women live in cells off the gallery. When asked if she will comply with regulations she answers 'Certainly'. She is accommodating. Despite this, a chain is 'put round her leg and fastened to the floor' but as this was happening 'she was perfectly composed and did not seem to take any notice of it'. Margaret remains stoic. The Patient Admissions log records her admission. It was confirmed by the Committee on Saturday 12 August 1786.

Margaret Nicholson admitted to Bethlem Hospital,
11 August 1787, 'admitted on the Incurables'.

A year later, on 11 August 1787, the register shows that she was transferred to the part of the hospital demarcated for 'Incurables'. Ten years later, on 28 January 1797, one James Tilly Matthews is admitted to the hospital and taken to the men's quarters in the east wing. Later he, too, is transferred to a section for incurables.

Matthews and Nicholson will both spend a very long time at Bedlam and acquire a fame over the years which spreads well beyond the walls of the hospital. But there are no records of what they made of each other inside, either by getting wind of a reputation or after a snatched conversation in person. Men and women would not normally fraternise, but Margaret and James were sometimes given special treatment at the hospital. They both had axes to grind with the authorities. They both tried to maintain a dignified attitude and an intellectual appetite – they were interested in the world. They both believed people in positions of power were doing them down.

As those who raided her apartment had found, Nicholson had few possessions, and, once at Bedlam, 'accordingly [she] was bought two new shifts…a pair of shoes…a black quilted petticoat'.[8] She was scrutinised like a prize heifer, and she invited varied opinions. Her composure slipped again. A physician found her 'much convulsed and…as if she was making an effort to weep, saying at the same time, "Tears would give her relief!"' The physician in question decides she is 'deranged'.[9]

The *Belfast Evening Post* of 10 August 1786 reflects on her true state of mind: 'There are intervals when lunatics assume reason, and are capable of conversing with a seeming rationality; but when close questioned as to a particular crime they may have committed, they wander into the wild labyrinths of distracted imagination, and discover their insanity. Such a one Margaret Nicholson appears to be.' The staff describe her in much the same way: as highly changeable, swinging from one state to the other. Thomas Monro is in overall charge of the place, the third generation of a great Scottish family of physicians. His appointment has been controversial. He acquired this coveted job without a formal election. The hospital's money is running out and his leadership and competence are continually questioned. He was a member of the Privy Council so he's already formed an opinion of Margaret. He offers her some help. Help not just with new clothes, but with getting her story out to the public.

Remarkably, she manages to have her petition published on 28 August 1786. Communicating to the outside world from her cell in Bedlam, she lays out the injustice of her situation on the page, adding a litany of new claims for good measure. She is descended from 'Boadicca the British Queen'... 'thy ministers are knaves – they grind the face of the poor with monstrous taxes – they have taxed our ribbons, our gloves, our pomatum, and our scented water... Make me a general, make me master of the ordnance, and I'll batter down the ramparts at Woolwich, and destroy the rotten stones of the Tower... Her whole fortune is at present a pair of Queen Elizabeth's stays, and a Queen Anne's farthing. Pity her sorrows, oh King. Dry up the tears of the afflicted Margaret Nicholson; the heir to the throne of the world wants bread.' 'So says Margaret Nicholson,' she adds, pleading her case in the third person for dramatic effect.[10]

The reporting continues apace over the next many decades, and diary stories provide updates on Margaret Nicholson's life on a regular basis. 'Peg', as she was nicknamed, was generally portrayed as a mad spinster. But the qualities of a romantic heroine show up in the report of her next bold move.

She's been allowed to walk about freely in the garden at Bedlam and used the time to make a plan. 'She formed, with some ingenuity, a ladder from...the handles of two brooms, connected by strong slips torn from her blanket,' says the *Kentish Gazette* of 30 November 1790. 'By this simple apparatus, she was able to ascend the wall, and thus made her escape without difficulty.' She escapes from the hospital.

Where can she go? She runs to the 'respectable' establishment run by her brother George, the publican at the Three Horseshoes on the Strand. The hospital quickly works out where she has gone, however, and staff hurry after her. She is apprehended. This time she apparently puts up a fight, but she is overpowered without much difficulty. More than a century later, just after the First World War, 'Madame M' will attempt something similar out of her Paris asylum, desperate to continue her guerrilla war against the conspiracy of substitute doubles. She will also be unsuccessful.

Margaret is back in Bedlam for the long haul. So how did a self-effacing and diligent woman come to believe not just that she'd been cheated out of a fortune, but that the throne of England was rightfully hers? Where did the delusion come from?

Parish records in the market town of Stokesley, near Stockton-on-Tees, in north Yorkshire show that Margaret Nicholson was born on 9 December 1745, making her forty when her name hits the papers for the first time. The reporters weren't far off with their biographical sketches. Her father, Thomas Nicholson, was a barber, and Margaret was the fourth child born to him and his wife Anne.[11]

In the year of her birth, episodes of anti-Catholic unrest connected to the Jacobin uprising in France disturbed Stokesley. A mob pulled down and burned a mass house in the town. John Wesley preached at Stokesley many times between 1752 and 1790 in the Methodist chapel built for him in the town. The mainstay of Stokesley's economy was the manufacture of linen from flax, and the associated spinning and weaving. When Margaret was growing up the work was only partially mechanised and still largely a cottage industry. The waste from the linen trade went to the mill to make paper.

Authentic Memoirs admits rather snootily that she was literate, and got her impressive needlework skills 'from the indulgence of her parents', receiving an 'education something superior to that usually given to the daughters of ordinary tradesmen in that part of the country... She was taught to read, write and work at her needle...in the latter she was perfectly adept, working with that delicacy and skill as to have been able to earn a comfortable subsistence.' There was a local charitable school. She may have developed her abilities with a parish apprenticeship in household service. Later, like 'Madame M', she is a lady of letters, and her writing has a vivid quality, confidently organised by themes and images, even though the letters make almost no logical or grammatical sense at all. She consumes literature enthusiastically. In one letter she quotes Viola in *Twelfth Night*, and William Knollys, 8th Earl of Banbury, describes her being given 'dictionaries to read' while chained to the floor at Bedlam.[12] Margaret's brother George says 'she employed herself in reading Milton's Paradise Lost and such high styled Books'.[13]

George came up to London from Stokesley at some point, too. Evidently a barber's wages couldn't support a large brood at home in Yorkshire, and even if Margaret's mother Anne supplemented the family income by taking in work for the linen trade, as many women did, both Margaret and her brother were required to be self-sufficient. They saw greater opportunity down in London. George did well. The Three Horseshoes thrived under him.

At the age of twelve, maybe thirteen, Margaret became a maid, and worked in a few notable households including that of Sir John Seabright, before she was employed by Lord Coventry. No one recorded any mention of mental illness.

Authentic Memoirs contains additional observations about her character which confuse the generally accepted picture of her as a hardworking servant. One source says she's on the make, as those young girls are, with a 'boldness, cunning and intrepid address, seldom to be found at a more advanced period of life'. The volume suggests she got ideas above her station from an 'appearance of probable promotion' with the Boothbys, evidently a household of particularly high status. This 'kindled the latent sparks of pride which lay hid in the recesses of her heart'. Pride, 'ruinous pride', is her downfall. It leads to 'insolent superiority' with fellow servants. She thinks she's better than them. The brother she ran to also tells tales on her. George Nicholson apparently told Fisk that his sister 'had been insane for some years, and that her insanity was occasioned by pride'. Here is another person ensnared in a cautionary tale.

And then, in the very final chapter of his pamphlet, Margaret's landlord Mr Fisk drops a sensational allegation. Stop press: he's had the chance to look over 'several particulars concerning Margaret...lately communicated to me'. He doesn't say what these particulars are – he's just the messenger – but 'Mr Paul [a mutual acquaintance of Nicholson and Fisk], Mrs Fiske, and some others think differently' about this woman and her so-called madness. She's not delusional at all. The word is she 'exercised great cunning and dexterity' in the 'science' of 'swindling'. She's after financial support from a 'generous' king. She's been trying to trick our beneficent monarch into giving her money. It's worked, hasn't it? She's got new clothes and a roof over her head for free.

Every time we think we're getting to know Margaret, building a picture of her, the elements of the photofit are mixed up again. The records are full of hearsay which has become established fact simply by endless circles of repetition, one source quoting a rumour back at another, round and round. Are people using Margaret to make money, even pulling her into their own paranoid conspiracies? If Margaret is a master swindler, could her history be something else altogether?

There are gaps. She moves from household to household but only a few of the most recent families employing her are named. No one mentions her religion, or lack of it. There is Margaret Nicholson in the records who married a Richard Wilson on 5 September 1773, in Kirkby, Cleveland, two miles from Stokesley. Our Margaret would have been twenty-seven. A Richard Wilson, born in 1736, died at Stokesley in 1822, aged eighty-six. Another Richard Wilson died in Stokesley in 1784. And another on 27 April 1777 in Stainton. But there will be Margaret Nicholsons in the area, it's a local family name, and there's no suggestion that our Margaret was ever married, or that she left London in the 1770s. Still we're seeing doubles of Margaret and a phantom husband, parallel lives.

Fisk was experienced in peddling sensational allegations for profit. In 1781 after being falsely accused of forgery by a tenant he published *The Case of Jonathan Fiske, bookseller: tried and honourably acquitted at the sessions of the Old Bailey, held in June 1781, upon the*

*infamous prosecution of Patrick Roche Farrill, for forgery: with anecdotes
of the prosecutor and his adultress confederate, Alice Harriot Herbert who
cohabits with him.*[14] He reserves the haughtiest contempt for Alice
the 'adultress confederate'.

Most likely Margaret was in service for all these years in
London. There are moments when we can confidently place her
in a certain household, getting on with her work.

One such moment places her at the centre of a compromising
event, in the house where she was living and working. This time
it's the *Scots Magazine* dropping a shock allegation into its copy.
There is a witness. The paper lines up the details and charges
the innuendo: 'Margaret Nicholson lived some years ago with
a lady of quality in Brudenel Street, as her own servant. Her
master's valet-de-chambre paid his addresses to her.' Margaret
had a man coming to call at night. You wouldn't have guessed
what was going on all initially. She thought she was clever enough
to keep the visits secret, 'her conduct before the family was very
reserved; but one of the family happening to remain up after the
rest were a-bed, in walking upstairs softly, at a late hour, surprised
the valet-de-chambre going out of her bed-room'.[15]

The family member was, apparently, shocked. It hadn't crossed
any of their minds for a moment that she had it in her, the *London
Gazette* says, because until that evening they had considered her
'reserved and thoughtful cast; seldom subject to the livelier sal-
lies of mirth; this constraint of temper was considered by her
fellow servants as prudery'. The family couldn't have imagined
'he had any prospect of success with her'. These are weapons'
grade backhanded compliments. She's far too straightlaced and
boring to be guilty, but that just makes it all the more shocking

that she is. The household source, whoever it was, could not resist a good gossip. 'On such a discovery as this, everyone knows how anxious the discoverer is to unburthen his mind… and next morning the servants were entertaining themselves at the expense of the reserved, as they called her, prude; the news soon reached the mistresses [sic] ears.' The unmistakable whiff of a sex scandal.

Margaret and the valet were both dismissed. Together they looked for a new place to live, and found one, where they remained for some time. But they had to leave that one, too, and sought a third, and this is where things really took a turn for the worse for Margaret. The valet cut his losses and deserted her. And then he chased the money, paying 'his addresses to a person who had some property; whom he married and then left the place he shared with Margaret to take an Inn on the Western Road'.

Margaret had been viewed (generally) as hardworking and disciplined. Now she was an outcast, sexually available, promiscuous. The only evidence: a fleeting glimpse of a man outside her room after dark. The dignity her tireless domestic work had earned was gone in a casual whisper. There were more whispers that she had become pregnant and that the baby had been removed from her shortly after birth. She was humiliated, shamed.

According to the tabloid editorial of the *London Gazette* Margaret hid away from other people. The journalist offers an explanation for a mental imbalance like hers: 'Intense thought upon one object debilitates the mind; and with a temper already prone to melancholy, an accumulation of thought and distress must encrease intense thinking, which cannot but produce paroxisms of madness.' Here was the old complaint rearing its head. At this point in time, melancholy is a sentimental frame for the sad story of a woman driven mad by heartbreak. But it also refers us back to very practical advice in Robert Burton's masterwork on the subject. The guidance comes in handy again after more than one hundred and fifty years: 'Society and variety are necessary to remove the ill consequences of melancholy; neither of these it appears she sought for; even her brother acknowledged that she seldom called on him.' It appears she cut herself off even from the only family member available to her. *The Plot Investigated* picks up the romantic angle on her predicament. Other papers joined in and ran with new details about what happened next. 'After this she stopped looking for positions as a servant and started earning her living through the needle.' The *Scots Magazine* chips in: 'From that time she abandoned herself to solitude; and hence, perhaps, was planted the root of her disorder.'

Margaret had supported herself in skilled service from the age of twelve but when we catch up with her next she is scratching around for piecemeal work. Without a position, she's getting together a living taking in sewing. She has skills, but this was an overcrowded and poorly paid market; a hand-to-mouth existence.

The *London Gazette* embellishes an update on Nicholson's sad case with a sharp reminder of just how easy it was to slip out of 'good society' after only a couple of instances of bad luck or a bad personal decision or two: 'It is no secret how many thousands of women are in want of bread, who strive to live by the needle; therefore we must infer that her mode of living must be penurious and low, the want of nourishment, with the attendant anxiety on this, must increase that mental debility which is the result of melancholy; therefore the effusion of a mind under these circumstances, must be out of the control of reason.'

Who is to blame for her reversal of fortune? It's hard to say. There's nothing she can do about a lover who chooses to reject her for someone richer.

And so Margaret begins to construct an alternative version of the events that had led her to this dire situation. She is, in fact, a person of high birth. Accordingly, she deserves courtesy, dignity, even admiration. And there is an inheritance to go with this birthright which she must fight for. There is even a crown. The 'monarch' keeping her place is an imposter. He must do the right thing and give her back what is rightfully hers, or she will have no option but to take more drastic action to right the wrongs and restore her reputation.

The delusion gives her an itinerary. It brings her to St James's Palace with a blank petition in one hand and a small, blunt dessert knife in the other.

The story of Margaret Nicholson becomes an important one in the history of delusions. Her 'grandiose' delusion bumps her up the social ladder above the king himself. It also turns her into a cause célèbre. Prospective visitors express horror at what she did that day at St James's, but they still want to meet her. She is interviewed by doctors and members of the public while incarcerated, and their wide-eyed appraisals invite us into her intimate world. The world of the asylum.

'Peg' Nicholson will be an object of curiosity at Bedlam for four decades. Many journalists call on her over the years, sometimes bringing artists along to draw her likeness. Close attention is paid to the way she dresses as well, her features exaggerated to the point of caricature.

Like Spira's before her, Nicholson's delusion is turned again and again into propaganda. She is brought into the debate over the state of the mental hospitals in Britain. In a Rowlandson caricature in 1793 called 'A Peep into Bethlehem' she appears as a wild-eyed grotesque crowned with straw and clutching fistfuls of the same in her outstretched hands. In another cartoon she is Charles James Fox in drag attacking the king. A portrait etching in an oval frame for the cover of the *New Lady's Magazine* is positively glamorous, with pearls and ribbons in her hair. But the most surprising is a late portrait. This is Margaret Nicholson in old age by John Thomas Smith, a painter, engraver and Keeper of Prints for the British Museum, as well as a mentor to a young John Constable and a long-time acquaintance of William Blake. Smith was a gossip and gadfly, in close touch with the artistic and literary life of London for over sixty years.[16] He was known for his unsentimental approach, capturing his subjects with what could be 'malicious candour and vivid detail'. His portrait of Nicholson shows a woman with a deep well of experience behind the eyes. She fixes the viewer with a penetrating stare from under a mob cap and headscarf.

Bedlam's myriad visitors exemplify the eighteenth century's curiosity about the 'mad'. Some are family to the residents, like James Tilly Matthews's loyal relatives, some foreign dignitaries in London, some simply day-trippers who treat Bedlam as an entertainment much as they would a zoo. The governors outlaw this, but, as the years roll on, Margaret is still the character people want to see more than anyone else.

Nicholson was officially insane but she was elevated to a 'personality cult'. Something about her story, her version of reality, and her belief in what she was due, spoke to people. They identified with her and indulged her grandiosity.

The German novelist Marie Sophie von La Roche got close access to Margaret Nicholson. She was a great traveller and kept a detailed diary. The prelude to the encounter was a tour of London when she taken right up the entrance to St James's Palace and shown the very place where 'the mad Nicholson woman made an attempt on the King's life'.

Von La Roche then pays a visit to Bedlam, which she calls a 'shrine of pilgrimage' for Germans. The great scale of the building and the imposing statuary at the entrance impresses her. She finds the place surprisingly calm inside, the cells relatively well appointed, and she is impressed with Monro, the doctor in charge. She is taken to see the star of the hospital: "'And now," said the supervisor, door key in hand, "I will show you Mistress Nicholson."'[17] Von La Roche gives a vivid account of the reveal: 'I shuddered at seeing a person with murderous instincts,' she says, but she is presented with a woman 'tidily attired, her hat upon her head with gloves and book in hand, stood up at sight of us, and fixed her horrible grey eyes wildly upon us'.

Like James Tilly Matthews a few years later, Margaret was given certain privileges not afforded to other delusional patients, including pen and paper with which to express herself. During her audience our traveller von La Roche witnesses a touching and intimate scene between the patient and an inspector. The official notices pens lying on the floor and Nicholson says they are useless. Nicholson is writing a letter to a royal. She wants to marry the Prince of Wales. She lifts the page, revealing elegant handwriting, but she is not happy with the composition. 'See here, the first lines were good, but I cannot let the Prince see the rest,' she says. The inspector offers to get her new pens. Von La Roche watches on as 'the sad woman thanked the inspector'. The inspector asks her if she has enough to read. She shows him the few pages she has left, and he says he will bring more books. Nicholson goes back to her current tome: 'It was Shakespeare she was reading so intently,' says von La Roche.

Another traveller, Jacques de Cambry, paying a visit from pre-revolutionary France, provided an additional snapshot of Nicholson in Bedlam two years later, in 1788. His point was political: this would-be regicide was being dealt with very differently from the man who tried to kill Louis XVI. Louis's assailant was considered sane and shown no mercy. Just like von La Roche, he finds Nicholson well turned out in a black hat. She is now reading *The Merry Wives of Windsor*.[18]

Nicholson was still a voracious reader of the canon. Perhaps Samuel Richardson's popular novel of 1740, *Pamela*, was brought to her at some point. The work was already a classic, and boasts a heroine, a servant, who ends up marrying her master. *Pamela* represents an era in which social mobility is possible, even for women. Potential is recognised and the virtuous and talented can overcome class barriers. Professional roles remained out of reach but making a good marriage was a perfectly legitimate route and used a woman's natural abilities. A romantic scandal could still nullify hard-earned respectability for ever. Nicholson knew this only too well.

By now the dilapidation at Bedlam was accelerating. In 1814, when Nicholson had been in the institution for nearly thirty years, Edward Wakefield, a Quaker and leading advocate of 'lunacy' reform, visited the hospital several times. He finds it failing to respect the guidelines of a progressive 'moral treatment': patients are not appropriately categorised, and are all bundled in together, regardless of the level of violence they display.

Wakefield's reports of incurable patients at Bedlam still in chains leads a House of Commons Select Committee to look into major reform. The findings in 1816 result in Thomas Monro's resignation. He was found 'wanting in humanity' towards his patients. The same report forces John Haslam out. James Tilly Matthews's evidence of abuses at the hospital carried weight despite all those years of argument about whether or not he was sane. In another ironic reversal, it is Thomas Monro who was asked to give an opinion on George III's mental illness. After his fall, Monro begins a new life as a major art collector and patron, establishing a prominent artists' circle in his London townhouse, which included J. M. W. Turner. In his later years Monro reportedly suffered from a delusion of his own. He claimed that he could raise people from the dead, and send others to the underworld.

With the old regime booted out, a new Bedlam hospital is eventually commissioned and the old one left to sink into the 'town ditch' rubbish tip on which it was built. Margaret Nicholson and the other Bedlam patients are transferred from the Moorfields site to the new one on St George's Fields in Lambeth in August 1815. They travel in a convoy of hackney coaches. James Tilly Matthews is somewhere among their number.

An anonymous insider gives an account of Bedlam, written when Margaret has been inside for thirty-six years. She is still denying she ever meant to hurt the king, and insists she always held him in high esteem. She says she only went to see him in person because she had known him as a child. Nicholson never renounced her royal connections. She is granted yet more privileged living arrangements because, even after all these years, she 'never evinced any prominent symptoms of insanity', and is 'tranquil and calm' just as the people she lodged with observed all those years ago. Now she is even allowed to live apart from all the other criminal inpatients 'in a ward appropriated as a nursery for the aged and infirm, such as are quiet and harmless. She enjoys a good state of health, is regular, cleanly, and attentive to her little concerns, and is desirous to render herself useful, such as her great age will permit.'

During her later years she apparently develops an aversion to bread, but is allowed gingerbread as a replacement, which she likes, and snuff, which she adores: 'her favourite luxury of which she takes a great quantity'.

Eventually she becomes totally deaf and can rarely be persuaded to speak, but is said to be in good health and spirits right up until her death.[19]

On the evening of 15 May 1800 there was another attack on George III. James Hadfield fired a pistol at the king at the Theatre Royal Drury Lane. He was brought to Bedlam. Did Nicholson ever know Hadfield? Did James Tilly Matthews? Probably not. The governors at Bedlam built a wing for the 'criminally insane' after another regicide attempt in 1816.

Margaret Nicholson died on 14 May 1828 after forty-two years in Bedlam. She outlived George III by five years and was convinced to the day she died that she was the rightful owner of half of England.

The subversive potential of Nicholson's case was noticed by a young poet named Percy Bysshe Shelley. His *Posthumous Fragments of Margaret Nicholson* is published in 1810, the same year as John Haslam's book about James Tilly Matthews and the Air Loom. Shelley's work is a collection of poetry subtitled: 'Being Poems found amongst the Papers of that Noted Female who attempted the Life of the King in 1786'. The pamphlet is a playful hoax. The fragments purport to be Nicholson's scribblings and ephemera found in her apartment among her personal effects. These fragments are in fact fictional creations of Shelley, and they include letters in which Nicholson lays out her claims to royal status. If it's a game, it's a serious one. Shelley attacks the British monarchy dressed up as Margaret Nicholson far more savagely than Margaret ever did with her blunt dessert knife. He calls the institution oppressive, and highlights the mad injustice of primogeniture.

Nicholson's encounter with the King of England took place just a few years before the revolution that swept across France. Her belief in her right to the throne was deluded, but it was also disruptive. Her delusion was a personal power grab, but it was also a symbolic act critiquing a society rigged in favour of the few. This was a demand for attention and influence made by an ordinary woman, a needleworker, the daughter of a barber from north Yorkshire, the sister of a publican. Absolute monarchs beware.

The papers settle on the cause of Margaret Nicholson's delusion, somewhere between love and ambition. We can try to piece together a more accurate photofit of the woman with cuttings. In the end, though, we can't see Margaret, or trace the line of her biography, through all the conflicting reports, rehashings and gossip. But we can listen to the delusion itself; listen to everything Margaret says. This is the clearest communiqué we have. The rest is noise. She's telling us how to treat her. She's telling us about her hopes and fears, about injustice, wretched bad luck, a thankless working life. That's it. That's the point of a delusion – it's as clear as a bell, when everything else is confused, contradictory or opaque.

Louis Capet Being Welcomed to Hades, Villeneuve, c.1793.

CHAPTER VII

The Clockmaker Who
Lost His Head

1793, the Marais, 3rd arrondissement, Paris, three miles north and over the river from the catacombs, a mile from the Prefecture of Police on the Île de la Cité, and just over a mile south-east of the Hôtel de Biéville, a celebrated clockmaker is at work in his studio. This workshop is one of many in the crowded street dedicated to 'horologe': the art and science of watches and clocks. Today, though, he is not working on a decorative clock or a watch, but something else, something groundbreaking. Shouts from a revolutionary mob in the street don't distract him and he concentrates on what's in front of him. Watch wheels and copper and steel plates lie scattered along the bench. He adds to the intricate construction of moving parts, adjusts something, measures, tests. This is a prototype. A prototype of a device, still hypothetical, that will work for ever without an outside energy source. He hasn't slept much for weeks but he's still going.

The 'Marais' refers to the marshland on which the district was built. The area has come up in the world and now it's the choice for aristocrats, with its rue du Temple serving as the main artery up to the Temple complex itself. The 'Temple' of the religious order of Knights Templar who founded a fortress here. King Louis XVI was held at the Temple until his recent execution, on

21 January 1793. (It was Louis XVI who survived an attempted regicide a few years previously, just before Margaret Nicholson's assault on George III with the butter knife.) It was from here, in fact, that Louis and his entourage left to travel to 'the machine' in the Place de la Révolution. The clockmaker heard the procession. Back to work.

Louis XVI's patronage has seen the golden age of French watch and clockmaking, and an increase in clock production, as well as all manner of associated mechanical discoveries. The manufacture of objects that record the time has evolved into a complex science, and art. There's a demand for ever more accurate timekeeping, particularly for maritime and astronomical use. This clockmaker has skilfully adopted the technology of the pendulum and the spiral balance spring, items invented in the seventeenth century by a Dutchman named Christian Huygens (curiously enough, this was the father of the poet Constantijn Huygens, author of *A Costly Folly*, which satirised 'glass delusion' by depicting a subject so terrified of smashing himself that he took to his bed).

Decorative clocks made in the Marais are known for their fashionable ornamentation, inspired by recent classical excavations, such as the lyre or urn. Glass-encased sculptures referencing Greek or Roman mythology are another speciality. The scientific clocks proudly show off their mechanical craftsmanship, boasting a new precision and even portability, in plainer designs. Many specialisms are involved in putting these timepieces together. French clockmakers are governed by the guilds who require them to employ independent craftsmen for each element – cabinet-makers, for example; stonemasons, or guilders – although the clockmaker retains control of the final product. Our clockmaker has built clocks to order for the wealthy of Paris, some of whom have recently been arrested, so he hears. He wonders if they will share the awful fate of the king. Back to work, back to work.

Our clockmaker has kept abreast of the recent scientific discoveries. He rubbernecked at demonstrations of the same unseen forces that captivated James Tilly Matthews. (Matthews is currently imprisoned in the Hôtel de Biéville across the city in the 9th arrondissement, the Air Loom just beginning to take shape in his imagination.) As a young man, our clockmaker heard people talk about Mesmer's apartments and the marvels of animal magnetism. The perpetual motion machine facing him is a natural extension of his expertise. Like a clock it also deals in time, but if he can make it work, it will have the upper hand over time's constraints. The concept of perpetual motion brings to mind astrological cycles and orbits and the horoscopes and futurology connected to them. The mechanics of the guillotine have shown how easy death can be. There is a kind of futureproofing in his endeavour, reminiscent of Burton's endless revisions of his *Anatomy*.

People have been trying to create a perpetual motion machine since medieval times, and the clockmaker has been infatuated with the idea since childhood. 'Cox's timepiece' inspired many young minds. It was developed in the 1760s by British jeweller James Cox, alongside John Joseph Merlin. Cox was a Walloon from Liège, a Freemason and inventor working in London. Their invention harnessed atmospheric pressure, via a mercury barometer, to keep the mainspring coiled, and brought to market a clock that didn't require winding. 'Cox's Timepiece' turns out not to be an example of true perpetual motion, however. In 1775 the Académie des Sciences in Paris declared the idea impossible and said in a statement that it would 'no longer accept or deal with proposals' after a steady stream of attempts that came to nothing.

The exhausted clockmaker has not given up, though. Not at all. As he continues to tinker with the calibration of his prototype he begins thinking again about the workings of that other new device he caught sight of recently through the crowds. The 'louisette', named after Dr Louis who helped bring it into service, the machine that will soon be known as the 'guillotine'. This machine is never mass-produced and, just like the finest clocks, requires a whole team of carpenters, metalworkers and blacksmiths to construct. To calculate the rate of the fall, the weight of the steel blade was measured against the 'mouton', the metal weight attached to it. That would tell you the speed of the decapitation. Our clockmaker didn't see the blade fall in the Place de la Révolution because of the throng, but he saw the blood meander across the cobbles around people's feet. He also saw the leather basket waiting for the head. He's back to work on his motion. But then he stops to really think about the workings of this other machine. He is confused, disorientated. He doesn't feel like himself. Suddenly, he understands.

*

In September 1793, at the height of the revolutionary Terror, Philippe Pinel, aged forty-four, took up his post at Bicêtre asylum and set to work categorising the 'known occasional causes' of the two hundred patients in his care. The principal categories were:

Domestic; Misfortune; Love; Religion or Fanaticism and Events connected to the Revolution.[1]

It was during this time that Pinel came across a number of patients with a belief about themselves which was so fanciful and yet so graphically macabre that it demanded special attention. Their heads, they said, had been cut off by the guillotine.

One such case at Bicêtre which struck Pinel as particularly interesting was that of a certain clockmaker from Paris. The man was obsessed with the idea of perpetual motion. Pinel theorised that this enthusiasm was related to the 'influence of revolutionary disturbances'. These disturbances had made their way into his mind and now 'his imagination was greatly heated, his sleep was interrupted, and, at length, a complete derangement of the understanding took place'. His family had sent him to Hôtel Dieu, and he was then transferred to Bicêtre.

Pinel outlines the details of his patient's primary delusion:

His loss of reason was marked by a most striking feature: he fancied that he had lost his head on the scaffold, that it had been thrown indiscriminately among the heads of many other victims, that...the judges having repented of the cruel sentence, had ordered those heads to be restored to their respective owners and placed upon their respective shoulders, but that, in consequence of an unfortunate mistake, the gentleman who had the management of that business had placed upon his shoulders the head of one of his companions in misfortune. The idea of this change of head occupied his thoughts night and day, which determined his relations to send him to Hotel Dieu.[2]

After listing to him speak, Pinel swiftly came to the conclusion that the clockmaker's delusion and others in a similar vein were a reaction to trauma. Pinel's decision to devote himself to this new discipline of psychiatry, and to what he called 'moral treatment', was whispered to have been inspired by the death of a friend in 1783 'who went insane through excessive love of glory' and died because 'all pharmaceutic remedies' failed to save him.[3] The baffling suicide drove him to his vocation, they said. He was interested in the origins of a seemingly mysterious imbalance.

Certainly, the guillotine would have traumatised anyone in or around Paris at the time of the Terror. Its development was a process of trial and error. On 17 April 1792, a herd of sheep were decapitated at Bicêtre by a brand new 'execution machine'. Pinel watched the practice runs. He then watched trials of the machine on human corpses. In fact he knew its inventor personally, as they were both involved in the debate over the soundness of ideas around 'animal magnetism' around the same time.[4] The mechanised blade severed approximately three thousand heads between March 1793 and August 1794 and inevitably the technology of the guillotine infiltrated the public imagination. Even if a person escaped its action, it made death so close you could taste it and the escape hard to believe in. In 1793 an unknown artist published a satirical engraving entitled 'Louis Arrives in Hades', which shows the French king and his court arriving in the Underworld after their executions. They are carrying on as normal, bowing, laughing, chatting and seemingly oblivious to the fact that they are carrying their own severed heads in their hands.

Pinel describes the clockmaker's behaviour on an ordinary day:

Nothing could equal his outrageousness and noisy bursts of jovial humour. He sang, shouted, and danced. And since his maniacal insanity entailed no act of violence he was allowed to go about the hospital freely in order to expend his tumultuous effervescence. 'Look at these teeth' he would cry 'mine were exceedingly handsome, but these are rotten. My mouth was healthy, this one is diseased. What a difference between this hair and mine before my head was changed.'

This is a scene from a drunken burlesque. Inside the world of his delusion, however, the clockmaker's concerns are perfectly logical. He has the wrong head so he has someone else's teeth. No wonder it would occupy his thoughts and require urgent attention. Pinel is not jumping in with restraints or attempted cures. He is indulging him, trying to understand.

Hailing from provincial Languedoc, Pinel studied at medical schools in Toulouse and Montpelier, and then went to Paris where he was professionally overlooked as something of an outsider. Pinel failed to get one of the grants going for poor students at the Faculty of Medicine and instead took a job at one of the private medical institutions which were springing up to take in the swelling numbers of mentally ill. He spent many years slogging away in these low-profile sanitoria, away from the high offices of national medical institutions. It looked as though his career would be stymied by his 'lack of fortune'.[5]

Pinel's nephew said that his uncle suffered from a chronic stammer that turned public speaking into a daily battle. He never overcame it and was consequently quietly spoken, even withdrawn in company. Since this nephew was by all accounts devoted to him and had no apparent motive to denigrate him, this detail goes some way to humanising the legendary Father of Psychiatry, as he became. Here's a man who found public speaking arduous and stumbled when he did talk.

On 16 March 1790, a law was passed which banned arbitrary internment of 'lunatics' based on just a single letter ordering it, and meant that they had to be seen and assessed by a doctor. This change was what got the Marquis de Sade out of Charenton, another Paris asylum, after his first incarceration there. The earthquake of the Revolution had shaken the impenetrable walls of the medical establishment and its hierarchies, and for the first time a person like Pinel could get in through the cracks. The profession's old guard had lost influence and several of Pinel's friends were now advising the new government. Pinel saw his opportunity, and his pioneering spirit took him quickly to the top at Bicêtre. There he formulated his thinking about delusions as phenomena related to human sensitivity, whose causes were to be found in ordinary life.

Bicêtre was developed as a hospital on the site of an old monastery just south of Paris in the middle of the seventeenth century on the instructions of 'Sun King' Louis XIV. The Marquis de Sade was in residence for three years before he was carted off to Charenton, and, like Bedlam in London, it doubled for decades as a prison, clearing the city's beggars and prostitutes from the streets.

Bicêtre must have seemed something like a medieval vision of hell when Pinel arrived. The official history is that he set about encouraging freer conversations with patients that were designed to wander off-topic, and that he observed a strictly non-violent policy, banning bleeding and replacing it with careful listening and detailed note-taking. It was Pinel who set out to dismantle the link between mental illness and demonic possession. He grasped that a mental disturbance might have a physiological cause, but it could also be a product of social and psychological stresses and therefore required psychological, or 'moral', treatment as it came to be known.

If Pinel's hunch was right about trauma and the clockmaker we can start to see how the delusion might operate, how it's calibrated, and what it might offer him. It's certainly given him a job to do. He must sort out the administrative error of the mixed-up heads and a kind of 'normal life' can resume. Getting his head and teeth back will keep him busy, and keep him out of the general confusion.

The clockmaker doesn't remain so fanciful in his approach to life. He is violent, just as the 'Napoleons' who will arrive at Charenton asylum in a few years could be. He smashes up or otherwise destroys anything within his reach, making close confinement necessary.

Although his belief about his head doesn't change, his behaviour does improve. The aggressive outbursts stop and he is allowed to walk freely again in the inner court at the hospital.

And then the other obsession returns. Something has been ticking away at the back of his mind. 'The idea of the perpetual motion frequently recurred to him in the midst of his wanderings; and he chalked on all the doors and walls as he passed, the various designs by which his wondrous piece of mechanism was to be constructed.' Pinel was gripped. He became more involved in the clockmaker's daily life.

The cure, Pinel decided, was likely to come from encouraging his patient's enthusiasm to finish the perpetual motion project. 'His friends were, accordingly, requested to send him his tools, with materials to work upon, and other requisites, such as plates of copper and steel, watch weels, &c. The governor permitted him to set up a work bench in his apartment. His zeal was now redoubled. His whole attention was now riveted upon his favourite pursuit. He forgot his meals.' There was a setback. 'After about a month's labour, which he sustained with a constancy that deserved better success, our artist began to think that he had followed a false route. He broke into a thousand fragments the piece of machinery which he had fabricated at so much expense of time, and thought, and labour; entered on the construction of another plan.'

He wasn't beaten. Pinel observed him as he slaved away for another fortnight, putting all the parts back together. He got them working and, eventually, he 'fancied that he saw a perfect harmony amongst them'. All eyes were on the machine and its creator. 'The whole was now finally adjusted – his anxiety was indescribable – motion succeeded.' It seemed as though he'd achieved the impossible. Could it be true? Pinel describes the scene of excitement: 'It continued for some time – and he supposed it capable of continuing forever. He was elevated to the highest pitch of enjoyment and triumph, and ran as quick as lightning into the interior of the hospital, crying out like another Archimedes, "At length I have solved this famous problem, which has puzzled so many men famous for their wisdom and talents."'

Success, vindication, elation. The world would be at his feet! We can only imagine the scenes at Bicêtre.

This great victory over the limitations of time and energy was short-lived: 'grievous to say, he was disconcerted in the midst of his triumph. The wheels stopped! The perpetual motion ceased! His intoxication of joy was succeeded by disappointment and confusion. But, to avoid a humiliating and mortifying confession, he declared that he could easily remove the impediment, but tired of that kind of employment, that he was determined for the future to devote his whole time and attention to his business.' The clockmaker didn't accept defeat. He simply had more work to do on a specific area. It wasn't that the machine could work, it was that it had to work. It was the carefully balanced mechanism of his mind keeping time over all the noise and chaos around him.

Now Pinel attempted an old-fashioned 'ruse'. He would enter part way in the delusion in the hope of bringing his patient back with him to reality. He'd practised with something similar when he devised a fake trial and 'pardoned' one of the many patients who believed they were heading for the guillotine. This time, he co-opted another patient of a 'gay and facetious humour' to play a part. This patient was instructed to interview the clockmaker, and to bring the conversation around to the miracle of St Denis, who had famously carried his head around in his hands after it had been chopped off. The clockmaker confirmed that the scenario was feasible. The actor-patient laughed at this and ridiculed him: 'A madman you are, for how could Saint Denis kiss his own head? With his heel?'[6] Presented with that one simple, irrefutable fact of anatomy, the clockmaker 'retired confused amidst the peals of laughter, which were provoked at his expense', and he 'never afterwards mentioned the exchange of his head'. Apparently, close attention to his trade for some months 'completed the restoration of his intellect. He was sent to his family in perfect health; and has for more than five years now, pursued his business without experiencing a relapse.'

'Pious frauds' as they were sometimes called, or little white lies, to trick a person out of a delusion have been around since people first started trying to cure them. Reports of the easy success of a ruse are also found all the way back in Burton's classical examples and are impossible to substantiate. Burton cites the melancholic from Siena who was scared to urinate in case he caused a biblical deluge, for example. In his case, a physician set a fire and the melancholic acted on impulse and put the fire out the quickest way he could by peeing on it. He was cured when a great flood didn't ensue. It's the same tactic used by the court doctor who started a fire to trick an unnamed royal out of his glass delusion, asking him why he hadn't smashed when he banged on the door trying to get away from the flames?

Pinel's use of the ruse at Bicêtre in the 1790s shows that some of his treatments were far from cutting edge. The counterpoint view on Pinel's regime is most savagely represented by philosopher Paul-Michel Foucault in his 1961 essay 'The Birth of the Asylum' in *Madness and Civilization*. According to Foucault, medieval treatments were simply replaced by other subtler means of control. They may have been old methods, but Pinel saw that playing tricks on the delusional could be more successful at influencing a person's thinking than authoritarian methods. Pinel wanted to cure delusions. This was new and ruses at Bicêtre were in the service of that objective.

In an editorial in the *Gazette de santé* before the Revolution he mentions a story which drew him to ruses as a way of speaking

to an imaginative illness. This was a young priest who had play-
fully consulted both a fortune teller and a reader of horoscopes
and been told by *both* he would die at twenty-five. In an echo of
Robert Burton's story, 'after the double prediction his 'exalted
imagination' caused the 'most violent alarm in his soul' and he
became 'withered and desiccated'. But when his twenty-fifth
birthday came and went, he was completely cured.[7] The date
confronted him with reality and allowed him to resolve things for
himself. The ethical problems with ruses, principally around the
question of consent, were another issue and only appeared more
and more obvious as the decades passed. A ruse may have cured
the clockmaker, but a few old tricks were no match for the scale of
the delusions inspired by the Revolution. They kept on coming.

The guillotine continued to appear in the delusional imagination of the French nation. There were cases of women at
Salpêtrière admitted in the early 1800s who shared their fears
of being guillotined. Pierre-Jean Laujon was the son of a famous
popular songwriter, who fled the country when the revolutionary
dramas began. Laujon had fought as a counter-revolutionary
but was arrested at the Swiss border, taken to Paris and incarcerated, eventually at Charenton. An unnamed actress later went
to Charenton to see a production of Molière's *Le Dépit amoureux*
staged by the Marquis de Sade, who had been in residence there
since 1803. She remembered the scene as being 'gaily performed'
by the 'madman', Laujon, playing the valet Mascarille, but the
doctor's report in the ledger at Charenton describes someone
more obviously troubled: 'He has the most bizarre ideas, he
thinks that he has been beheaded and that his head is still in
England, that someone has probably given him another in its

place; to replace missing teeth, he constantly wears pieces of cork in his mouth.'[8]

There's a memorable image of decapitation lodged in the mind of one of Charles Dickens's characters in *David Copperfield*. Mr Dick is a wise fool, working obsessively on his own memorial but distracted by his other obsession with Charles I's head. There is often the feeling that fictional characters with delusions are trying to communicate something uncomfortable about the world, something which the people around them don't want to hear. They themselves, on the other hand, have found peace, like Cervantes's *Don Quixote* further back in the previous century, happily tilting at windmills, believing that he's a chivalric hero, while everyone laughs at him.

Trauma can produce delusions that are protective, like our clockmaker's, or that are pitch-perfect metaphors for experiences. In the aftermath of the First World War, delusions caused by 'shell shock', or what we would now call post-traumatic stress disorder, were common. In Mary Borden's memoir of working as a field nurse in France during the Great War, *The Forbidden Zone*, she remembers: 'My body rattled and jerked like a machine out of order. I was awake now, and I seemed to be breaking to pieces.'[9] The belief that you had lost a limb when it was in fact still attached was reported in military hospitals across Europe.

The other side of our clockmaker's delusion – the quest to achieve perpetual motion – also pops up again, with an interesting twentieth-century spin, in *Cause for Alarm*, Eric Ambler's spy novel of 1938 set in contemporary fascist Italy. Our expat fugitive hero has just lost his job as an armaments manufacturer and takes refuge in the mountains with a mathematical genius.

The professor has fallen foul of Mussolini by refusing to sign a declaration that fascism is the 'sacred religion'. What's ruined him professionally, it turns out, is a swivel-eyed delusion regarding perpetual motion. The professor has become convinced that the laws of thermodynamics are based on a gigantic misconception and that 'science was nothing but a house of cards'. 'I knew for certain that he was insane,' our hero concludes before he dashes off.

The suggestion, though, is that a false belief like this might be a perceptive response to a mad world. In 1960 the psychotherapist R. D. Laing will tell us about a little girl he met who said she had an unexploded atom bomb in her stomach and draw the same conclusion. It is easy to dismiss our crackpot inventor as an amusing anecdote. His beliefs were supposedly reversed with a simple trick. But we might consider the possibility that he had a tighter grasp on the true reality of the Terror than some others who stayed calm and composed as they got on with their lives.

Modern theories of thermodynamics proved perpetual motion to be impossible. We understand the clockmaker's attempt to create a more reliable, and predictable, future, and what a crushing blow it must have been to have his machine fail. All the same, it's hard to resist the idea that he might have had another go at it in a few years.

Unfinished Portrait of General Bonaparte,
Jacques-Louis David, *c.*1798.

CHAPTER VIII

Napoleon and 'Delusions of Grandeur'

IO June 1831. It's nearly three years since Bedlam registered the death of one of its longest-term and most notorious patients, Margaret Nicholson. James Tilly Matthews has been in his grave in London for fifteen years. At Charenton asylum in Paris, a patient is admitted. A doctor makes notes about the man in the 'Register of Medical Observations':

> The first day we found him dressed elegantly, head held high, with a proud, haughty air; his tone was that of command, and his least gestures indicated power and authority. He soon informed us that he was the Emperor of France, with millions in riches, that Louis Philippe was his chancellor, etc. Then he...pompously recited verses of his own commission, in which he allocated kingdoms, settled the affairs of Belgium and Poland, etc.[1]

There's a stagey drama to his entrance, like a guest at a costume party and some of the same black comedy as with the glass king. To the staff at Charenton, though, this man means trouble. He's not the first Napoleon to knock on the door of an asylum in Paris. And Napoleon has been dead for ten years.

We can picture this 'Napoleon' instantly, in his three-dimensional glory: the iconic bicorn hat, frock coat and proud gaze, short stature and puffed up, hands behind his back. His very name personified power and ambition and was a shorthand to the staff at Charenton, as it still is. The situation deteriorates into something much more desperate. According to the physician's notes, this 'Napoleon' was violently disruptive:

> During the day he smashed everything because people would not obey his every order. He was calmed by a shower and then shut up in a cell. The next day we found him naked, having torn up everything, shouting, threatening, etc.

We notice a pitiful disconnect between the proud icon 'Napoleon' and this raging unclothed man in an asylum cell.

The phenomenon of duplicate 'Napoleons' is spreading. The First Emperor of France continues to show up to other hospitals with orders. Nine years later, as the real Napoleon's coffin is returned to the French capital, Dr Voisin, the head physician at the Bicêtre asylum, will note 'fourteen or fifteen emperors' in his hospital.[2] As with glass delusion centuries before, here are multiple individuals experiencing a delusion with the same characteristics. Research in the city archives by the historian Laure Murat found that 'delusions of grandeur' accounted for more than a quarter of mental disorder diagnoses in the register for the Greater Paris area for 1840.[3] And these delusional people are taking the shape of one grand figure above all the others: Napoleon Bonaparte.

The unshakeable belief that you are a person of great fame

or power, when patently you are not, is a 'delusion of grandeur', another category in the World Health Organization's taxonomy of delusions dating back to the sixties. It's the second most common delusion type after paranoid, persecutory delusions, and it's so fixed in the popular imagination that for many delusions of grandeur have come to represent the subject as a whole. This is probably in part owing to the splendid fancy-dress parade that accompanies the cases but also because grandiosity is often an attendant trait in people experiencing other primary delusions, as it was with 'Madame M' and James Tilly Matthews.[4] Grandiosity is subtly different from a delusion of grandeur. The grandiose are *related* to Boudicca, or the king, or they have a fortune owed to them, or they are at the centre of diplomatic relations during the French Revolution. With a delusion of grandeur you *are* Napoleon.

Delusions of grandeur are the simplest expression of how delusions seem to function; a primer in the help and protection they offer a person. And Napoleon is the poster boy for the whole operation. The Charenton Napoleon of 1831, and the fourteen or fifteen other emperors after him at Bicêtre, jump out of the dusty ledgers of multiple nineteenth-century establishments. The message couldn't be clearer. This is a declaration of importance. Here are the powerless asserting their authority. Like the Glass Men, they are telling the world how we should view them; how we should treat them, but they are shouting the instructions not whispering them. 'Respect me!' they cry. They are proactive, turning things around for themselves. It's a basic mechanism transporting people who are knocking along on the bottom to positions of great wealth and power.

Laure Murat even found evidence of a *woman* who believed herself to be Napoleon: a seventy-one-year-old admitted in June 1852, though it's not clear if she believed herself to be Napoleon I or III. 'She says she's Napoleon; she shouts, "Long Live Napoleon!"'[5] Such was the power of Napoleon to trigger delusional illness in the mid-nineteenth century. It's as disconcerting to count them all as it is to tally 'Madame M's substitute doubles. But these are not impersonations, they are not doppelgängers, they *are* Napoleon. Identities which feature in delusions of grandeur are unambiguous and unchanging like caricatures. Despite appearances, the individuals experiencing the delusion resist a simple interpretation, and we can lose them just as quickly as they arrived back into the throng or the gloomy backstreets.

Delusions of grandeur are perennial, but they have gone by different names over the millennia. The personas which crop up again and again to strut and fret on the stage are, predictably enough, the most powerful and influential figures of recorded history. No one is on record as being Louis Philippe – the so-called 'Citizen King' who took over in France a few years after Napoleon and was mocked for being utterly ineffectual. Delusional personas are from the landed nobility, or they own all of India, are Louis XIV, or Charlemagne; even Satan, and, not infrequently, Jesus.

This particular delusion goes back as far as the records. Robert Burton included a few choice examples of what we would now call delusions of grandeur in *The Anatomy of Melancholy* in 1621: a woman from Genoa, for example, who believed she was married to a king, thus making her queen. She conversed with this king, and when she saw bits of broken glass in the street, she said they were jewels that he had given her. A man from Cremona thought he was the pope and set to work appointing cardinals. Shakespeare is playful with grandiose delusions in *The Taming of the Shrew*. His drunk tinker, Christopher Sly, is found asleep, and on waking up he's tricked into thinking he's a nobleman.

By the time these nineteenth-century emperors were marching around, the belief that you were a world leader when you were not was already the quintessence of insanity. Thomas Hobbes wrote in 1651: 'If some man in Bedlam should entertain you with sober discourse; and you desire in taking leave, to know what he were, that you might another time require his civility; and he should tell you he were God the Father, I think you need expect no extravagant action for argument of his madness.'[6]

Napoleon is an archetype of political power, Christ of spiritual power. In Western Europe, people who believe that they are the Messiah have made themselves known for centuries. They were spotted in the seventeenth century by Hobbes proselytising in their cells at Bedlam, and spotted again visiting the Holy Land itself. According to the Israeli psychiatric doctors who have to triage them, a few tourists to the biblical sites still present as the Messiah each year. These 'Christs' are connected to what's become known as 'Jerusalem syndrome', a term coined in the 1930s by the Israeli psychiatrist Heinz Herman, where a person arriving in the city is overtaken by a state of high emotion and devotion, usually someone who is already an evangelical Christian.

Delusional 'Messiahs' arrived well before the formal description of the syndrome. A Frenchman, Simon Morin, is the most famous early case. He was burned at the stake in Paris in 1663 for his belief. The burning was in a public place, in the Paris of the 'Sun King', another absolute monarch whose confident vision was reflected in the new classical façades across the city. Morin's burning exposed the tensions beneath these new façades. The Jansenist movement inside the Catholic Church was threatening the order of things with their emphasis on predestination and grace. Descartes was writing stirring rationalist critiques – about the contemptible nonsense of glass delusion, for example. It's in this context that Simon Morin stands up and announces that he is the Messiah, and, according to his conviction about who he is, he makes the ultimate sacrifice. A 'delusion of grandeur' is often attached to a fearsome sense of responsibility, for saving the world in Simon's case. The roar of power starts to sound more anxious. The message has more complexity and nuance.

Delusions of grandeur are as old as the hills. Nevertheless, the sheer number of people exhibiting them in Paris at the beginning of the nineteenth century was remarkable. Physicians and the general public alike puzzled over the phenomenon. You could be forgiven for thinking that delusions of grandeur (discussed side by side with a new concept of 'monomania' where a single fixation existed in an otherwise unaffected mind) were at epidemic levels in the city.

Why did so many people believe they were Napoleon, specifically, at this place and time, and not some other dignitary or royal from the past or present? Why did so many people identify, not just with him, but *as* him? Becoming this one man like Charles VI and the others became glass. Napoleon Bonaparte is long dead by 1840. His rule ended in 1815 and his last breath was taken in exile on St Helena in 1821. These Napoleons hold on to their personas with an iron grip.

Doctors at Charenton were continuing to refine the new discipline of psychiatry, along with their peers at the other great nineteenth-century establishments: Bicêtre, Salpêtrière, Sainte-Anne and Hôtel Dieu. Jean-Étienne Esquirol became chief physician at Charenton in 1826. He was a protégé of Philippe Pinel, who died that year, and he seized on the Napoleons as an opportunity to advance his mentor's work. Under Esquirol's leadership the physicians at Charenton continued to challenge the association of delusions with moral weakness and the idea that a person experiencing one was a vessel for the devil. There were still people who came to the hospital claiming to be Satan, but this was a psychological phenomenon, not demonic possession.

Under Esquirol, these 'Napoleons' would not be contained and punished like criminals. They would be listened, to with a view to curing them. Asylums like Charenton were still formidable places, no question, but Esquirol and his colleagues followed Pinel's lead in a rigorous attention to note-taking and careful logging of symptoms. The startling numbers of Napoleons in the registers points to there being more of this kind of delusion per head of population in early nineteenth-century Paris than at any other point in history. At least part of the explanation for this, however, lies in the simple fact that they were keeping such extensive records.

By the time the fourteen or fifteen emperors turn up at Bicêtre in 1840, France is a country exhausted by decades of war, and the weak, charisma-free Louis Philippe is in charge.

The French had fallen a long way in a short time. A few years later, one of the contributors to the *Bibliothèque du médecin-praticien* went further by saying: 'we have no fear of being taxed with exaggeration in saying that that the madness of the age is pride' and that never before in history 'have we seen so many men make themselves out to be saviours, to have first-rank talents and abilities'.[7] The sin of pride had been identified as the culprit for mental instability for centuries. Wounded pride in a personal sense is something different, and it has also driven delusions under the radar and is thus harder to spot. Margaret Nicholson, James Tilly Matthews and Francis Spira were all influenced by it.

The people of France were looking for someone larger than life. Napoleon was a figure from a noble past. He was dead in the literal sense, but in another important sense he is even more alive, and necessary, than he had ever been. This is the archetypal self-made man. He came from the island of Corsica, not from Paris, and went on to command an empire. Napoleon represented the supreme triumph of perseverance over the accident of birth. The opposite to that lottery of primogeniture. His father was from a noble Italian family, but Napoleon spoke Spanish and was ridiculed for it in the mainland French military school he attended, as well as for his short stature. The battle for the new Republic brought him back from Corsica. He won promotion in 1793 after defeating the British at Toulon. Suspected of treason he was imprisoned for ten days, but unlike James Tilly Matthews he managed to negotiate himself back on side with the regime. When forces loyal to the king attempted to retain power in 1795 Napoleon was called in to stop the uprising. His prize was the badge of Commander of the Army of the

Interior, then his Josephine, then Italy, Austria, Egypt. The rest is history (until the defeat at Waterloo) but the takeover was improbable. You wouldn't have bet on it. Here was proof that magical thinking could work, and take anyone with enough grit to the world stage and then to its control room. He even managed to wrestle an accommodation with the Catholic Church in the Concordat with Pope Pius VII of 1801 by reaffirming its majority status among the other denominations. When the dust settled a few years after Waterloo the story that had survived about Napoleon was a simple one charting his domination of the Western world.

There's a timeless draw to any story of an individual rising from a modest start to great success and power. Napoleon was a costume that felt good on, offering protection from a wretched existence, even a reversal of fortune – his had flipped in the right direction. The revolutions in printing and pamphleteering allowed the personality cult of Napoleon to be disseminated in a mass-media explosion.

Esquirol had a theory. He saw the 'Napoleons' and the other 'grand figures' walking the corridors at Charenton as a product of the frenetic empire-building overseen by the first Emperor of France. This was a time when territories were being fought over, boundaries drawn and encroached upon, fixed national identities and power bases destabilised:

At that epoch, when [Napoleon] peopled Europe with new kings, there were in France many monomaniacs who thought themselves emperors or kings, empresses or Queens. The Spanish war, conscriptions, and our conquests and reversals also produced their mental maladies. How many persons, stricken with terror at the two invasions, remained mono-maniacs! Indeed, we now find in madhouses many people who consider themselves dauphins of France and destined to the throne![8]

In this treatise from 1820 Esquirol looks back on recent history and makes a clear link between delusions of grandeur and their empirical context, especially an unstable – unreliable – political structure. He refers to 'two invasions', Austria in 1805, and Russia in 1812, as well as to the Peninsular War with Spain to the south-west in 1808. Allegiances broke up and reformed with England, Prussia, Russia, with tides of attack and retreat on both fronts.

In line with Esquirol's analysis, the rise of the nation state across Europe coincided with a fair number of 'Empress Eugénies' and 'Emperor Wilhelm Is' and, during the geo-political creep of the Third Reich in the next century, it was not uncommon for women to claim Hitler was, for example, their brother. Margaret Nicholson was also an international sensation at this time. If Esquirol ever heard about her delusion in doctors' despatches across the Channel from Bedlam he might well have put it down to the pride of the British Empire.

The dream of the Revolution had become an endless night-mare by the time the 'Napoleons' arrived. According to Esquirol, people had continued to respond to each successive trauma with delusions: 'The influence of our political misfortunes has been so great, that I could give the history of our revolution from the taking of the Bastille to the last appearance of Bonaparte from that of certain insane persons whose insanity connects itself with the events which have [so marked] this long period of our history.'[9] Esquirol's take is that delusions are ground markers for the country.

Esquirol locates delusions in broad brush strokes of French history. There is one particular Napoleon delusion, a companion piece with a compelling twist on the original that is worth homing in on and considering in more detail. In the *Analectic Magazine*, an American digest of the best papers and articles of the day, there is an account of a European woman with a 'delusion of grandeur' who believed she was Napoleon's wife. A letter of 12 June 1817 by an unnamed correspondent reports the author's encounter with 'Madame Bonaparte'.[10] She is apparently experiencing her delusion while Napoleon is still alive in exile. The correspondent's gloss on the situation is that she was from a 'reputable family', but 'had an inordinate attachment to splendour and equipage; which the circumstances of her husband would not permit him to indulge... Her dress became highly expensive and fantastic: and she would take possession of any elegant carriage which she found drawn up at a neighbour's house; giving the coachman directions to drive up to some spacious abode, which she deemed her own. In one of these excursions she was driven to the lunatic asylum; and, rather against her will, detained there. It was, however, "her palace"... At this time Napoleon Bonaparte enjoyed the wealth and dignity of a powerful emperor; and who should be the husband of our lady, but the potent monarch of France, and the temporary creator of the destinies of Europe!' The eye-roll in the commentary makes fun of her claims. Her belief that she was empress was so strong that the commonly understood reality of the world and its property would adapt to fall in line with it. Then the story gets more complicated. She is not having a good time. She is in prison. 'She was his spouse, but confined in a castle. He was testing her affection for her. She wouldn't recognise her husband or daughters and dressed in a regal fashion even when confined.' Here is another delusion of recognition. She won't accept her children as her own. The author says they have in their possession a letter which the woman sent to Napoleon from 'Spain' on 26 March 1816. She actually wrote it in 1811. They transcribe 'Madame Bonaparte's words:

> – what can I say to you? A volume could not contain it – and yet my pen is mute; nor can my hand, my tremulous hand, retrace the great, the vast, the awful ideas that nearly over- power my imagination; nor engage in that converse sweet that is comprised in objects more minute. I certainly have caught the contagion, or mania of objects that surround me. I am bewildered. The sublime, the profound, the infinite; the burlesque and trifling; the tender and endearing; the repulsive and forbidden; sham quarrels, and checked, reconciliations: – grandeur, magnificence in prospect; real sufferings, indignities and respect, the sway of the hearts and affections of millions in submissive subjection to a small single control, &c. &c. – are so blended and confounded, that I can give no intelligible expression to any of them.

Her letter goes on to address her imprisonment. She wants to be free, but she doesn't understand why she isn't. Who is supposed to make the next move? Is he waiting for her to do anything?

This is a perverse version of true love where a mystery motive is keeping them apart. It's moving to read a first-hand account of a 'delusion of grandeur' that doesn't come filtered through a doctor. She is eloquent, existential, mysterious. How to treat something like this? What is going on?

At Charenton, Esquirol was interested in causes primarily as a means of improving treatments. One of the therapies tried at the hospital before his time was theatre. Charenton's director at the time, Abbé de Coulmier, had been interested in psychotherapy through art and for several years prisoners performed alongside the guards. There was a packed bill of productions including works by Molière, with the playwright Pierre de Marivaux as master of ceremonies. The Marquis de Sade was in Charenton for two stints, once in 1789 and for a longer stretch in 1803 until his death in 1814, and he was allowed to write and direct plays as part of the theatre-as-therapy initiative. Audiences were invited in from outside to watch and Hippolyte de Colins, a former cavalry officer, got a seat for one particular performance in 1812. After the show de Colins witnessed one of the actors, a famous dancer called Trénitz, raging to the others at having to take off his costume. He was dressed up as a king for the play. Those in authority concluded the plays might in fact make delusions worse rather than better and in 1813 they were banned.[11] So was there anything new that would help with a 'delusion of grandeur'?

There were always the old tricks. Ruses were one of the only methods in the professional tool kit still reporting some success. When a captain of the dragoons experiencing the delusion that he was the Napoleon turned up at a home for the 'mentally deranged', Dr Leblond, who ran the place with his father, attempted a 'pious fraud'. Leblond was in a standoff with the Emperor of France and entered just a few steps into the world of the delusion in the hope of negotiating a truce. He was met with outrage:

> It is surely an indignity to treat Emperor Napoleon in this way!' he declared to the doctor. 'Those frightful valets bound me – I intend to have them shot.' To which Leblond calmly replied: 'Yes, you are indeed emperor Napoleon, but Napoleon on Saint Helena'. On hearing these words the madman fell silent, then began repeating 'Saint Helena, Saint Helena.' He then asked to be unbound and kept his promise to remain calm until he was freed.[12]

The resolution is suspiciously fast. But Leblond is respectful, and he was apparently successful in de-escalating the situation. Physicians have always understood that simply throwing logic or even conclusive evidence at a person experiencing a delusion does not loosen the grip of their belief. A willingness to engage with the alternative logic of the delusion, to meet it at least some of the way, can be effective.

There's something about the pomposity and theatricality of a person with a delusion of grandeur that makes a ruse, which is like a parlour game, a particularly sympathetic match. There's an elaborate ruse to cure a 'delusion of grandeur' in the 1921 Pirandello play *Henry IV*. An unnamed nobleman falls off his horse while playing Henry IV for an annual carnival. He wakes up believing that he *is* Henry, an eleventh-century emperor. He's lives like this for twenty years, his family maintaining the charade in a villa made to operate like the court. Now a doctor arrives planning to shock 'Henry' out of his delusion. Hanging on the wall is a portrait of the nobleman wearing his carnival costume posing next to the woman he was in love with at the time. This woman has returned, with her grown-up daughter (a lookalike) and fiancé, to witness the ruse. The portrait is swapped for one of the daughter and her fiancé, and the rival is stabbed. The twist comes in Act II when the nobleman admits he has known the truth for eight years but would rather stay as Henry than live in the twentieth century. A king's identity is a place of relative safety. On the other side of the coin is our glass king Charles VI who shatters under the weight of his crown. Either way, they've attained their power by accident of birth or strength of belief, rather than by divine right.

There's an example of an American hospital hosting an old-fashioned ruse even later in the twentieth century, with serious ethical repercussions. In 1959 the psychologist Milton Rokeach brought together three men claiming to be Jesus in the Ypsilanti State Hospital in Michigan hoping to cure them of their delusion. In his 1964 book *The Three Christs of Ypsilanti* he wrote about the inspiration for this daring experiment.[13] A commentary by Voltaire caught his attention first. By the by, in his notes Voltaire mentioned the most notorious case of a delusion of grandeur, and heresy, the 'unfortunate madman Simon Morin', who said that he was 'incorporated with Jesus Christ'.[14] Voltaire relays the story of Morin at an asylum meeting quite by chance someone else claiming to be 'God the Father' and being 'so struck with the folly of his companion that he acknowledged his own, and appeared, for a time, to have recovered his senses'. It didn't save Morin. The authorities found an apology to God tucked into one of his stockings, but he was nonetheless burned alive in 1663 for his heretical claims. But it planted the seed of an idea for Rokeach.

Rokeach then came across an article by the psychoanalyst Robert Lindler in *Harper's Magazine* called 'The Jet-Propelled Couch' about a man who believed he was living half his life on another planet.[15] Lindler's line was that when a person enters – invades – another person's delusion, the person with the delusion is forced to give way. He gives an example of the tactic: 'imagining an encounter between two victims of, let us say, the Napoleonic delusion. The conviction that each is the real Napoleon must be called into question by the presence of the other, and it is not unusual for one to surrender, in whole or in part.'

Confront to cure is the theory. Lindler had more for Rokeach in the article: an extraordinary encounter that demonstrated the method. At a psychiatric sanitorium in Maryland, Lindler reports, were two middle-aged, paranoid women who both said they were Mary Mother of God. They were reportedly 'mild mannered', both Catholics, and of 'similar socio-economic level'. The two deluded patients began talking:

> Before long each revealed to the other her 'secret' identity. What followed was most instructive. The first, our 'oldest' patient, received the information with visible perturbation and an immediate reaction of startle. 'Why you can't be, my dear,' she said, 'you must be crazy. I am the Mother of God.'

The new patient regarded her companion sorrowfully, and, in a voice resonant with pity, said, 'I'm afraid it's you who are mixed up; I am Mary.' There followed a brief but polite argument which I was restrained from interfering with by my older and more experienced colleague, who bade me merely to listen and observe. After a while, the argument ceased, and there followed a long silence during which the antagonists inspected each other warily. Finally, the older patient beckoned to the doctor standing with me. 'Dr S.,' she asked, 'what was the name of our blessed Mary's mother?'

'I think it was Anne,' he replied.

At once, this patient turned to the other, her face glowing and her eyes shining. 'If you're Mary,' she declared, 'I must be Anne, your mother.' And the two women embraced.

According to Lindler, the woman who surrendered her Mother of God delusion 'responded to treatment' and was discharged.

Rokeach conceived an idea for another ruse, a formal experiment this time, to disrupt delusional belief systems. He brought the three men claiming to be Jesus together and made them interact while he observed. First, they engaged in spiritual debate over who was the holier, then there was a physical fight, and finally they each explained away the others' delusions – they were mentally disabled, or dead, or it was the 'machines' inside them producing the fraudulent claims. Each said the others were 'crazy' or 'duped'. Then Rokeach and his hospital team got more involved in the action, sending letters to the three men written in the guise of people who might influence their thinking. After two years Rokeach hadn't seen any significant change.

In the afterword to the 1981 edition of his book about the experiment Rokeach expressed regret. He apologised for the manipulation without consent. 'I really had no right, even in the name of science, to play God, and interfere round-the-clock with their daily lives.' It had he said, cured him of his own 'God-like delusion' of manipulation and control.[16]

He had become 'increasingly uncomfortable with the ethics of confronting people with delusions'. He could now see that something was going on in a delusion of grandeur that made a lot of sense: 'Striving for morality and competence, are universal human motives.' Napoleon wasn't famous for his morality, unlike some of the most enduringly popular figures like Mary, but he knows what he's doing and he knows why he's doing it. He represents absolute confidence in his actions and an unblinking self-justification.

Rokeach's work also made links between 'baseless' delusions and the rickety foundations so many of us use to construct our identities. *My belief is my identity*, say the 'Christs' the 'Marys' or the 'Napoleons'. The world may not accept this; they are not literally the people they say they are, but what can a person do to prove their essential self anyway? What can anyone else do to disprove it? Stories about successful ruses in the past seem more and more dubious in the light of more recent experiments like Rokeach's. Challenging a person with a delusion just made them freeze and put up the defences. They weren't vulnerable to logical proof or hard physical evidence. Rokeach admitted that self-confrontation was much more effective but that relied on a person wanting to step out of a protective persona that was serving them well. Even Lindler's story of the two 'Marys' who changed after meeting each other at the hospital was a subtle one. They didn't snap out of it. They adapted their delusions to accommodate each other and they stayed in a delusional world. Aside from the ethical issue, the major one with the confront-to-cure ruse was that it didn't work in the long term. Ruses fell firmly out of favour as a psychiatric treatment.

No ruse worked on Esquirol's 1831 'Napoleon', if they attempted one at all. Esquirol and colleagues had progressive ambitions, but treatments didn't always get very far. The 1831 'Napoleon' was discharged three months later, 'not entirely cured'.[17] The cause of this man's delusion retreats away from us as he marches out of the records and back to the streets of Paris.

Alongside any psychological drivers, there may be an unde-tected biological dimension to delusions of grandeur from the past. Parallel roots in genetic predisposition or neurological damage. Scientists looking for consistent patterns of brain disease or damage with delusions are homing in on the right hemisphere of the brain.[18] Lesions in patients with grandiose delusions are typically of the right frontal lobe. It's the right hemisphere which deals with perception of self and other; the left hemisphere can be thought of as a creative narrator. The theory goes that right-lobe lesions stop a person regulating people, places or body parts in relation to self and so the left brain steps in with a false or excessive explanation to make sense of the confusion: you're made of glass, your husband has been substituted for a double, you are Napoleon. The left side makes sense of the faulty perception on the right and invents a narrative. Delusions of persecution and guilt are typically associated with lesions on the left side in the temporal lobe.

Delusions of grandeur are also common as a secondary mania, in conjunction with diseases like Huntingdon's, Parkinson's and Wilson's as well as with abuse of substances which disrupt neurotransmitter function, and with 'bipolar' depression. Religious practice can prepare the ground for them. Thomas Jefferson University in Philadelphia conducted brain imaging studies of people in a state of devotion. The limbic system, the centre of emotions, shows much higher activity when a person is in this state, and the frontal lobes, which play the key role in keeping it calm, shut down. This may go some way to explaining how people visiting the holy sites come to believe that they are the Messiah.

Do all the Napoleons have right-lobe lesions? Our interpretation of delusions adapts with each new insight into the organic brain, but the balance between biological and psychological is unclear.

We still don't really know what to do with people presenting with a delusion of grandeur. There is an essential simplicity to a Napoleon delusion. Being Napoleon confers protection, influence and hope for the future. There is more than a little logic in the 'madness' of trying on the persona for size. It is a copycat power grab for anyone who feels at the mercy of the world. So far, so understandable.[19]

But then we remember 'Madame Bonaparte's letter to Napoleon and it's not so simple. It's worth replaying her words. They offer a rare inside story on a delusion of grandeur. 'I am bewildered,' she says. 'The sublime, the profound, the infinite; the burlesque and trifling; the tender and endearing; the repulsive and forbidden; sham quarrels, and checked, reconciliations:– grandeur, magnificence in prospect; real sufferings, indignities and respect, the sway of the hearts and affections of millions in submissive subjection to a small single control, &c. &c.– are so blended and confounded, that I can give no intelligible expression to any of them.' 'Madame Bonaparte' communicates the complexity and ambiguity that can underpin a 'delusion of grandeur'. There's a mystery to people, a melancholy to use the old word, that they don't give up easily, even if they are presenting us with a caricature. Do we leave them in peace, as the three 'Christs' should have been left? Or do we listen more closely?

'Boulevard de Strasbourg, Corsets, Paris', Eugène
Atget, gelatin silver print from glass negative, 1912.

CHAPTER IX

'Madame X', Professor Cotard and 'Walking Corpse Syndrome'

1874, Vanves, outside Paris, a forty-three-year-old woman turns back into the room away from the view of the lovely purpose-built gardens that surround the private mental institution in which she lives. These are the green spaces in which melancholy patients are invited to lose themselves. She informs her physician that she has just had a strange and revelatory physical experience. Her physician is the neurologist and psychiatrist Jules Cotard. He is ten years her junior, and he listens intently as she describes an extraordinary feeling, something electrical, like lightning, which ran all the way up her back to her head, accompanied by a noise which she thought would split her in two along the spine. The event has set off a chain of thoughts and revelations about herself, her body and spirit, which has led inexorably to the conviction that she is dead.

Six years later, in 1880, Cotard presents a new illness to the medical community. He believes it to be a type of melancholia. His lecture draws upon numerous audiences with 'Madame X'. It is delivered before the Société Médico-Psychologique in Paris on 28 June of that year, to be published in the *Annales Médico-Psychologiques* in September. His patient has expanded on her extraordinary experience and the sensations which came with it and he is now in a position to describe a new subcategory of the delusion type that alters a person's view of their own body:[1]

> Dr Jules Falret and I have had the opportunity to observe a patient suffering from a specific form of hypochondriacal delusion. Madame X complained that 'she did not have a brain, nerves, chest, stomach or guts; all she had left was the skin and bones of her disorganised body' (her own words).

Cotard continues, briefing the room on the key aspects to her delusion. The belief 'seemed to have developed as an extension of earlier metaphysical ideas that her own soul, God and the devil did not exist either. She believed that, because of the state of her body, she did not need to eat, could no longer die a natural death and the only means to put an end to her life was to burn her alive. Consequently, Madame X constantly asked to be burnt alive and indeed attempted on various occasions to set fire to herself.' This horrifying image calls to mind Simon Morin, burned alive for calling himself the Messiah, or indeed countless 'heretics' from previous centuries. In this case, it's not a fate ordered by the authorities; it's one requested by the person who would go on the fire.

Cotard then asks his audience to go back with him over the evolution of this nihilistic attitude: 'In 1874, at the time of her admission, the patient was 43 years old and had been ill for 2½ years. Her condition had started one day when she heard that "crackling of her back extending up to her head". After this macabre experience she became prey to constant weariness and anxiety, she felt "like a lost soul" and repeatedly sought the help of priests and doctors. She tried to kill herself on a number of occasions and she was admitted to Vanves.' Vanves is one of many independently run asylums that have sprung up in recent years. It's overseen by Pierre and Jules Falret. Jules has also observed 'Madame X'.

Cotard's observations point to a psychological crisis. The patient expressed profound guilt and shame at her own past behaviour. But it's also a spiritual one. 'Believing she was damned, her religious beliefs encouraged her to blame herself for all kinds of sins, particularly for having done wrong during her first communion.' Cotard shows us a woman organising all the scattered pieces of her past into a strict religious framework. She brings to mind Francis Spira and his eternal punishment. 'Because her life was but a string of lies and crimes, she fully deserved having been punished by God to suffer forever.' Then came the realisation that she was dead. In her account, with its lightning bolt, this has the transcendent quality of a religious epiphany.

Shortly after her admission, she had the sudden experience that she could 'understand the truth'. In other words, says the doctor, the 'delusions of negation' were now firmly fixed in place. She is unbalanced, violent (she calls acts of violence 'acts of truth').

We want to know what has led 'Madame X' to this point. But 'Madame X' is a person hidden behind a pseudonym, like Capgras's 'Madame M', and Burton behind 'Democritus Junior'. She's a headline for Cotard's new psychological phenomenon and he has to protect her anonymity. The only thing we know for sure about 'Madame X' is that her belief she's already dead is unshakable.

Cotard continues with his lecture: 'After a few months, she started to improve from her anxious melancholia. She was calmer, but in her daily interactions she became malicious and sardonic. Her delusions, however, did not change, and she continued to complain that she did not have a brain, nerves or guts, that eating was a useless torture, and that death by fire was the only solution to her woes. On examination, she showed bilateral reduced sensitivity to pain in most areas of her body: for example, she would not show any reaction when pricked by a pin. At the same time, touch, and the special sense modalities seemed all within normal range.'

Like Joseph Capgras testing 'Madame M's knee reflexes, Cotard is open to the possibility of biological reasons behind 'Madame X's delusion. She is apparently numb to pain when jabbed with a pin. Is this brain injury or neurological disease? Is it related to the tingling up her back? We can't know that part of the story. Cotard leaves the physical examinations there. His hunch is that this is really a psychological story.

His patient has annihilated herself. She is absent. A blank. Her denial of the physical world reminds us again of Francis Spira and his delusion of despair three hundred years earlier. Cotard will write about it as a sort of reverse grandiosity. There is a profound sense of alienation in her delusion, but although it's passive in nature it demands the same level of attention. Both individuals experience a revelation, both refuse food. Spira is plagued by guilt about his legal double dealing in pursuit of riches. 'Madame X' apparently believes that she has not led a good and honest life. She is tortured by this certain knowledge. As with Spira, there's the suggestion of secrets. If people knew what she had done, they would judge her and condemn her.

It's a puzzle. Cotard doesn't say whether she ever divulged what she had done that was so wrong, or who she had lied to. Perhaps this was paranoia and she hadn't done anything worthy of blame at all. Where did the delusion come from? Is there any logic to it? 'Madame X' is a negative impression of a woman. She's retreated into the dark spaces. Cotard followed her there, over the course of his conversations with her, trying to shed some light and expose her, give her some definition.

The principal types of delusion have remained consistent over the centuries and the recurring themes predate the psychiatric textbooks. There is a recorded case of a living death, a total self-annihilation, many years before Cotard brought 'Madame X' to the world and formally described a 'délire de négation'. This example comes to us via *The Anatomy of Melancholy* of 1621 by courtesy of Robert Burton's raid on the notes of a sixteenth-century physician in the Dutch Republic, Petrus Forestus, known as the 'Dutch Hippocrates'. Forestus tells the story of a melancholy patient who thinks that he is dead and refuses to eat anything. There was a ruse. According to Forestus, the doctor asked an associate to pretend to be another corpse. He then put his stooge 'in a chest like a dead man, by his bedside, and made him rear himself a little, and eat: the melancholy man asked the counterfeit whether dead men used to eat meat. He told them yay; and did eat likewise and was cured.'[2] Again, this feels suspiciously rapid. The farce of a 'dead body' sitting up in a coffin to shock another person out of a delusion has a ghoulish entertainment value, but physicians were prone to talking up the success of ruses for their own reputational ends and it's quite possible the trick was no more successful ultimately than the one pulled on the three 'Christs' or Pirandello's Henry IV. These interventions were dubious long before they were considered unethical.

People with the delusion that they are dead won't be easily revived, by any means, and walking corpses recur again and again. A persecuted melancholic in Tscherning's poem 'Melanchey redet selber' of 1655 regularly fancies that he has died, or that he is hanging on the gallows.

A key shock factor with the delusion that you're dead is the relative youth of the people experiencing it. They are typically somewhere in the middle of their lives, like 'Madame X'. In 1788 the Swiss–French natural scientist Charles Bonnet encountered an elderly patient, and in this account the delusion is a very different proposition.[3] This old woman apparently felt a draught and was paralysed down one side of her body. As per instructions, her daughters dressed her in a shroud and put her in a coffin. When she fell asleep they put her back in bed and gave her a powder containing opium, and this seemed to help, but the symptoms returned from time to time. This version of the delusion is more poignant, something like a memento mori.

Bonnet's patient was an outlier. Most of the already-dead are not very advanced in years. In his lecture of June 1880 presenting 'Madame X' to the world, Cotard produces five further sensational cases of a delusion of negation. These cases came to him via Jean-Étienne Esquirol, who had reported these examples of what he called 'demonomania': one woman believing that the devil inhabited her body, torturing her, and she could never die. The second was convinced that the devil had taken her body away, she was only a 'vision' who had been alive for thousands of years, and that a malignant being in the shape of a snake lived in her womb. A third believed a bad spirit had stolen her body and she had no blood. The fourth said that she was made out of the devil's skin and filled with snakes and toads; the fifth that she did not have a heart and as a consequence she, too, could not die. Cotard then cites two further cases in his lecture originally reported by François Leuret, a disciple of Esquirol. One man believed that he was damned and that he was a statue made out of imperishable flesh possessed by the devil. Like 'Madame X', he threatened to set fire to himself. The other, who was also damned and immortal, had a hole in his abdomen and was without a soul. He talks about the 'strange logic' and also the paradoxical nature of some of these cases of 'hypochondriacal' delusion, where patients say they are 'neither alive nor dead, or that they are alive-dead'. These patients are described as being unhappy in their delusions of immortality. It's a different kind of unhappiness from that connected with delusions of grandeur where people are frequently weighed down with heavy responsibility and sacrifice underneath their pompous confidence. Cotard mentions a man experiencing a hybrid of delusions. He 'believed he would never die because his body was protected by certain privileges awarded to him by Napoleon himself'. Cotard notes that 'hypochondria-cal' delusions, unlike anxious melancholia, seem to worsen over time. They tend not to end well. He is careful to separate the phenomenon from generic religious insanity.

Cotard continued feeling his way, trying to reveal his 'Madame X'. He developed his thinking further, publishing a paper in *Archives de Neurologie* in 1882, and entitling it 'Du délire de néga-tions'. The word *délire* as Cotard meant it carries a far heavier freight of meaning than the English word 'delusion'. Like the idea of melancholia, it is multi-dimensional and expansive. A 'délire' of 'négation' is complex, existing at the intersection of various complaints such as guilt and anxiety but they come together to represent something clear: complete and permanent self-annihilation. Cotard uses everything in his diagnostic repertoire to shed some light, to get through to her, maybe even to animate her.

As the character of 'Madame X' is styled by her doctor, Jules Cotard is himself reversioned through the mind of someone else. In his case it's a master impressionist of bourgeois Parisian society, Marcel Proust. Proust gives Jules Cotard a cameo appearance in the second volume of *À la recherche du temps perdu*, *Within a Budding Grove*. The novel was published in 1918, thirty years after Cotard's death. Cotard appears as the influential forward-thinker Professor 'Cottard', cunningly disguised with the addition of an extra 't' in his name.

Jules Cotard was well known to the Proust family. Marcel Proust's father, Dr Adrien Proust, was an eminent Parisian surgeon and a contemporary of Cotard at the École de Médecine. The junior Proust's life overlapped with Jules Cotard's for eighteen years, and so he was able to model the character of Cottard on a man who was not only an influential Parisian figure but someone he had observed at close quarters throughout his childhood.[4]

The real doctor disappears behind his literary persona but if Proust's version is anything like faithful to the living, breathing Cotard, he was greatly admired in his professional life for his vision and diagnostic skill. We first meet the fictional version in *À la recherche* when 'Cottard' is introduced as a prospective dinner guest of the central Swann family. He is portrayed as a man with a brilliant clinical touch, diffident but solid as a rock in his diagnostic instincts. In the book, the most intelligent of the younger doctors apparently says 'that if they themselves ever fell ill Cottard was the only one of the leading men to whom they would entrust their lives'.[5] But Proust does not hold back in presenting the doctor's flaws either:

> Cottard's hesitating manner, his excessive shyness and affability, had, in his young days, called down upon him endless taunts and sneers… Wherever he went…he would assume a repellent coldness, remain deliberately silent as long as possible, adopt a peremptory tone when he was obliged to speak, and never fail to say the most disagreeable things… Impassiveness was what he strove to attain, and even while visiting his hospital wards, when he allowed himself to utter one of those puns which left everyone, from the house physician to the most junior student, helpless with laughter, he would always make it without moving a muscle of his face, which was itself no longer recognisable now that he had shaved off his beard and moustache.[6]

There's more than a touch of the ridiculous about Proust's Cottard. He's trying his best to seem aloof and authoritative but can't pull it off. The students flock to him for advice but they find the man funnier than his jokes.

Jules Cotard was born at Issoudun in the Loire Valley in central France on 1 June 1840. His father was a bookseller and printer and, according to Antoine Ritti, the man who delivered his eulogy, a 'serious and reflective character' whose all-round, multi-discipline interests and education founded 'a philosophy which embraced science, societies and their developments, with an ethic which had, at its base, human nature'.[7] Cotard was an intern at Salpêtrière, where he first became interested in mental illness. Like his literary incarnation, he won prizes at medical school, and, just as Proust suggests, his colleagues came to him for consultations.

Like Capgras in the following century, Cotard is fresh from military service when he first meets the patient who will make his name. Their paths cross in 1874, only three years after the Franco-Prussian War, where he served as a physician in the infantry regiment in the French army. By 1871 he was back in Paris, treating the mentally ill, and, typically, also the poor, referred to him by the gendarmerie from all corners of the city. As with Joseph Capgras after the First World War, it is not difficult to imagine how a young doctor exposed to the traumas of physically and psychologically wounded soldiers might develop an enduring, even morbid, interest in the darker recesses of the human mind and the imaginative strategies that can develop in the dark.

The day Cotard sat down with 'Madame X' and heard about the lightning flash and cracking noise down her back, he had only just become the lead physician at Vanves. Vanves was a large private clinic with extensive gardens at a time when many European mental institutions were thinking about how to design the environment, inside and out, to be therapeutic. The German psychiatrist Maximilian Jacobi was working on this very question and visited Vanves as part of the whistle-stop tour for his report of 1841. He described how 'the views...from the windows, embrace flowerbeds and shrubberies, beautiful landscapes, or the grand park... Whilst the range of view from the windows of the restless and maniacal patients is narrow and uniform, that from the windows of the melancholy patients is wide and variegated, surprising and delighting the eyes by the multitude of beautiful objects presented to it.'[8] The view is designed to cater specifically for the 'melancholy' patients who are set apart from, and above, the others patients. A Tudor melancholic like Henry Percy wouldn't be out of place in this scene, posing in the natural landscape with a high-minded sensitivity to his surroundings.

Cotard is interested in his new variety of melancholy. While at Vanves he writes four papers on the subject, developing his thoughts further each time, and showing a deep and enduring curiosity.

Over the course of these papers, Cotard references other cases of negation he has come across in his clinical practice at Vanves in the 1860s, 1870s and 1880s. He identifies recurring patterns: 'If one asks them what their name is, they do not have one… They do not have an age. Where were they born? They were not born. Who were their father and mother? They have no father, mother and no children.' One man refuses to wear clothes 'because his whole body is nothing other than a large nut…he refuses to eat for he has no mouth; he refuses to walk for he has no legs.'[9] Five out of eight cases refused to eat, and many more cases feature bowels and digestive systems not working. 'Madame X', of course, reported having 'neither brains, nor nerves, nor chest, nor intestines'.

There were other cases that Cotard didn't know about. Dr Clouston's lecture at the Royal Edinburgh Asylum listed a variety of delusions reported by the women at the hospital. As well as describing the case of a woman who thought her legs were made of glass, he gave us other beliefs: 'being followed by the police', 'the soul being lost', 'that the head is severed from the body', 'having neither stomach nor brains', 'her children being killed' and 'being dead'.

'Madame X' was not alone in her delusion, but her experience of it was lonely. At Vanves although her view of the grounds from her room was designed to lift her melancholic spirits and improve her wellbeing, nothing slowed the inexorable deterioration of her condition. Her negation delusion had a powerful religious dimension and related sense of doom which did not fade.

Cotard was still asking questions about the nature of her delusion. What did it come from? Why was it so unstoppable? We might ask, too, why he was interested in her. They were from the same generation and that gave them certain defining events in common. In particular, everyone had just lived through the protracted end of the Franco-Prussian War. Cotard, as we have seen, had served as a regimental physician in the infantry all the way through. 'Madame X' was admitted to Vanves just three years after the Siege of Paris which forced the end of the conflict. 'Madame X' would have been in her late thirties, unlike 'Madame M' who was a baby during the terrible years of the blockade. The years of siege correspond with the onset of 'Madame X's delusion.

Andrew Hussey, historian of Paris's darkest chapters, writes of the truly gruesome reality of the food shortages for the people of Paris, resulting from Bismarck's attempt to force surrender: 'As the siege hardened,' Hussey writes, 'the most desperate among them took to digging up corpses in various cemeteries around the city, mincing the bones to make a thin sort of gruel, which offered little nutritional value but at least kept them warm.' This astonishingly gruesome scene turned out to be a prelude to the glamorous belle époque, just around the corner but unimaginable at this point. By mid-October fuel was low, and the trees on the Champs-Élysées and the other great boulevards were chopped down for firewood. All but the wealthiest went hungry. Hussey brings us the notes of the theatre censor Victor Hallays-Dabot. On 10 November, he noted that 'rats were being sold in Les Halles at 25 centimes each. An American, Wickham Hoffman, stationed at the American Legation, also recorded that dogs sold from 80 cents up, according to size and fat.' Brewery rats were 'a titbit'. On 5 January 1871, Bismarck began a relentless bombardment of Paris in which at least four hundred people were killed 'by the same model of Prussian gun that had been proudly displayed in the heart of the city at the exhibition of 1867'.[10] We don't know how close 'Madame X' got to the reality of starvation but even the luckier ones were plagued by survivor's guilt. She struggled for decades with the question of whether or not she was 'worthy' of food, or of anything for that matter. Everyone in the city was caught up in the trauma and humiliation. There was good reason for an unspoken mutual understanding between 'Madame X' and Jules Cotard.

There's another, more opaque dimension to the delusion: the question of the religious crisis that 'Madame X' described in detail to Cotard, which apparently set in several decades before the Siege. She specifically mentions 'having done wrong during her first communion', and the guilt she felt as a result. She directs Cotard to this particular event, singling it out for attention, suggesting there's an explanation to be found there. It's an enigmatic reference. What could she have done wrong at a first communion?

She makes her strong religious faith clear. She was born within a year of the July Revolution of 1830 which replaced King Charles X with his cousin Louis Philippe, a pale imitation of Napoleon by all accounts and not inspiring source material for delusions of grandeur. But she grew up in a time when the church was tolerated again, after its suppression during the Revolution, and charitable Catholic institutions and congregations were flourishing. The public relations around being a Catholic still had to be well managed, however. A person's first communion was a highly significant public occasion for a family, and especially meaningful at this point in time. In the early 1840s this would have taken place at twelve years old.

Manuals produced by the church taught children the protocol of their first communion and called it the most solemn and important event of a person's life. The severest spiritual penalties for any transgression were made clear in black and white. Extensive preparation was called for. Special clothes were bought or hired for the ceremony and girls would be expected to wear clean white dresses to reflect the purity of their soul. Confession was a prerequisite for receiving the Eucharist and a child would have been visiting the confessional for about five years before first communion, since the age of seven or so. It was sacrilege to receive the sacrament in a state of mortal sin, as was failing to observe a fast from both solids and liquids from midnight on the day of the reception.

On the big day, the communicant kneels at the altar, the girls covering their hands in a linen cloth to avoid touching the host. An altar boy holds a tray below in case any crumbs of the host fall from the priest's hands. The communicant is asked to reflect on what they, a miserable, poor sinner, are about to receive. This is the body of Christ. The communicant offers pleas, prayers and petitions for help to be worthy of the privilege. The extended family and wider community watch. A feast has been prepared at home where expensive presents await: a cross in ivory or gold, or rosary beads.

There are, then, many points in the proceedings when a girl could make a slip. So what did 'Madame X' do wrong? Perhaps she didn't fast properly. Or stumbled as she went to kneel at the altar. Perhaps a crumb of the host fell. Or her clothes were not perfect. Perhaps she hadn't confessed, or not to all her sins anyway. Whatever happened it was in front of the worldly authorities in her life, her parents and priest, as well as the divine.

Over the course of a confession to her physician decades later, 'Madame X' makes clear that, in her mind at least, this error turned her finest hour into a terrible humiliation and public shaming.

Whatever judgement she faced at home was complicated by the renewed scrutiny of the government of her family's Catholic beliefs. By the time her illness set in, the religious climate for Catholics was much more unsettled. The Third Republic introduced many anti-Catholic laws and clergy of the Roman church were seen as being associated with the old regime. Numerous laws were passed to weaken the church's hold on what was taught in schools. Republicans claimed that one of reasons the Germans had won the Franco-Prussian War was because of their superior education system. In 1879, priests were excluded from the boards of hospitals and charities. The anti-Catholic Jules Ferry Laws of 1882 were coming, and would mean that religious instruction was expressly forbidden.

Records don't show if 'Madame X' married or had children of her own. Jules Cotard did both and was by all accounts a devoted father.

Someone in 'Madame X's birth family had access to enough money to be able to send her to a private institution like the one at Vanves with ambitions to be a therapeutic environment. Putting relatives into private institutions is a popular nineteenth-century literary trope, and the problem of what to do with an individual who lived in their own private reality was widespread but also shrouded in secrecy and shame. Delusions, like other forms of 'madness', were still touched with ideas of devilry, or moral frailty.

We don't know what happened to 'Madame X' in the years before Vanves; the biography is a blank. Perhaps the best thing once again is to listen to the delusion itself. In the world she has created she is dead. Like Francis Spira, she has withdrawn from life. She is immune from prosecution for her actions or her character.

A person absenting themselves is a common feature of delusions. It's there by another name in the torpor and inertia of 'Madame M' and our glass king, as well as in Andy Lameijn's patient in Leiden who could go from being 'there' to being 'not there' at the flick of a switch. The records in Bedlam give us the account of Ellen Hamilton in 1892. She experienced paranoid delusions and claimed the people who were persecuting her down the telephone line 'had murdered her on six occasions'. She 'has no heart, no lungs, no body. Her brain had been cut out and she'd seen it lying about on the floor of the hospital…treatment with prolonged baths…still complaining that she was dead.'

Samuel Beckett wrote to the poet and critic Thomas MacGreevy about a visit to Bedlam in 1935, forty years after Ellen Hamilton's time there. He describes how he 'went round the wards for the first time, with scarcely any sense of horror, though I saw everything, from mild depression to profound dementia'. He seems to have kept himself at a distance from what he witnessed on that visit, but his later work suggests it influenced him profoundly.[11] Several of Beckett's characters talk after death, and with others the 'I' of the speaker disintegrates, resulting in a powerful sense of alienation. Beckett's 1946 short story 'The Calmative' opens with the narrator declaring: 'I don't know when I died.' The delusion of death makes itself useful in the twentieth century after the First World War. The person who thinks they are dead is the perfect nihilistic trope, just as the man who has lost his head was in France after the Revolution.

In a psychiatric context 'Cotard's syndrome' came to be read as an extension of severe depression; a person's explanation to themselves and other of their experiences of disassociation and alienation. In fiction walking corpses become vampire and zombie stories. There's a spin on the conceit. The alienation is not from the point of view of the person who believes they are dead, but from the mind of those who fear it as an external threat. 'Varney the Vampire' was the first popular image of the living dead, the star of a gothic horror serial within the 'Penny Dreadful' magazines of the 1840s, and he was the warm-up act for Bram Stoker's iconic creation Dracula in the novel of 1897. In the 1968 film *Night of the Living Dead*, director George Romero uses a cast of zombies to critique American society and the blind conformity of the population. Americans were in the middle of a war they couldn't win (Vietnam) and audiences packed into movie theatres to watch the 'Living Dead' destroy civilisation as we know it. The archetypal zombie originated in Haiti where a 'zombia' was a figure stuck between life and death. In this original version, they are not simply the living dead, they are substitutes for loved ones. You recognise them, but they are not the real person. They provoke the same sense of familiarity and strangeness as 'Madame M's doubles. There's a similar overlap in delusion stories. The boundaries between delusion types are permeable and different image systems often combine to express hybrid hopes and fears.

The living dead found their most shocking expression on film. Psychiatry, too, was exploring ideas of personal alienation and in 1960 R. D. Laing produced his landmark *The Divided Self*. It was a groundbreaking analysis of alienation as a psychological phenomenon. His central proposition was that mental disturbance came from the tension between the two personas inside us: our private identity and the self we present to the world. We derive our identity from others, and they from us, and if this transaction is insecure, crisis results. Laing was exploring how to treat his catatonic patients. They had completely withdrawn from the world. Some behaved as though they were already dead.

Laing attributes the catatonia in his patients to claustrophobic families and their demands. Here we're reminded of Andy Lameijn's reading of his patient who had been in an accident before the onset of his glass delusion. The parents, Lameijn surmised, had been overbearing as a result of the accident and this patient's particular spin on the delusion meant he could disappear at will. In his 1964 book *Sanity, Madness and the Family*, he talks about withdrawal, 'a strategy which a person invented in order to live in an unliveable situation'. In *The Politics of*

Experience and the Bird of Paradise from 1967 he said that breaking your connection to the world was not necessarily 'breakdown' but could instead be a 'break-through'. Being already dead was a response to interpersonal, social and existential factors, not just biological or neurological. It was a strategy. It made a lot of sense psychologically, and if a person was listened to, natural healing was possible without drugs or restraint.

Laing took Cotard's idea of a delusion of negation into the 1960s. He shares an account of his encounter with a young patient who, like everyone else at the time, was living through the Cold War. 'A little girl of seventeen in a mental hospital told me she was terrified because the Atom Bomb was inside her. That is a delusion. The statesmen of the world who boast and threaten that they have Doomsday weapons are far more dangerous, and far more estranged from "reality" than many of the people on whom the label psychotic is fixed.' The little girl has turned her body into a warning about the threat of nuclear annihilation. He asks that old question: who should we consider delusional and who rational?

'Cotard's syndrome' is rare, and, although persistent, it's not yet fully understood. There are neurological explanations in play now along with the psychological theories, traced once again to right temporal lesions in the brain. The syndrome is associated with a disconnection between the sensory areas of the brain and the limbic system, which is responsible for emotions and memory. This breaks a person's emotional relationship with the outside world and leads to feelings of unreality and delusions of death and negation. There's a link to migraines, too, as well as the use of LSD.

Neurologists continue to find out more about the organic causes of delusions, through ever more sophisticated diagnostic technologies like the MRI scanner. In 2007 a variety of receptor encephalitis caused by an ovarian tumour and linked to delusions was identified by Josep Dalmau at the University of Pennsylvania. The growth instigates an autoimmune attack, but there are particular brain cells that resemble the embryonic cells in the ovary and they are mistakenly treated as though they present a similar threat and attacked also. The resulting encephalitis inflames the right hemisphere and delusions result. An American journalist, Susannah Cahalan, wrote about her experience of these delusions caused by an ovarian tumour in 2009 when she was twenty-four. In *Brain on Fire: My Month of Madness*, she details her paranoia, believing that her partner was unfaithful, as well as numbness down the left side of her body echoing the experience Charles Bonnet's elderly Swiss woman who asked to be put in a coffin. She also had periods of catatonia. Practical tests showed that her spatial awareness was disordered and that the cause of the delusion was neurological not psychological. This type of delusion caused by an ovarian tumour is associated with grandiosity, and it can lead to violent hallucinations. Characteristic symptoms with this particular encephalitis include grunting and growling with convulsions and may even help to explain historic cases of 'demonic possession'.

With 'Madame X', as with our other subjects, there's the possibility of undiagnosed organic brain disease. Despite the continuing advances and shiny new technology, though, the brain and the mind remain profoundly mysterious and delusions elude a simple explanation of cause and effect. Psychological elements won't be easily disentangled from biological, and each delusion is the product of multiple, overlapping influences and inciting factors which remain opaque.

In 1985, the neurologist Oliver Sacks published his bestseller *The Man Who Mistook His Wife for a Hat*, a series of essays outlining the case histories of some of his patients. The title of the book refers to one patient in particular, a man who suffered from 'visual agnosia', another disorder of recognition. Sacks makes the case that classical neurology has traditionally neglected disorders of the right hemisphere in favour of the more easily demonstrable malfunctions of the left. The right brain 'controls the crucial powers of recognising reality which every living creature must have in order to survive', but they are strange and mysterious and 'somehow alien to the whole temper of neurology'.[12]

Despite the close-up view offered by MRI scanners, we are still in the foothills of understanding the mind. The Russian neurologist A. R. Luria, working in the earlier twentieth century, argued that an alternative neuroscience was needed: a 'personalistic' or 'romantic' science. This approach would be aligned closely with literary criticism and celebrate the literary qualities of case studies and the subjective and imaginative storytelling contained within them. In other words, we might get further in our understanding if we read an account like 'Madame X's as we read a fable. Sacks characterises his own subjects in *The Man Who Mistook His Wife for a Hat* as archetypes – heroes, victims, warriors. They 'hark back to an ancient tradition: to the nineteenth-century tradition of which Luria speaks; to the tradition of the first medical historian, Hippocrates; and to that prehistorical and universal tradition by which patients have always told their stories to doctors'. Sacks suggests we listen to the stories themselves. He condemns modern methods which 'could as well apply to a rat as a human being'. Today science is wary of the anecdotal and prefers to look instead to large quantitative analysis and control trials for its data. His plea is to 'restore the human subject at the centre – the suffering, afflicted, fighting, human subject'; to look for the person and the particularities of their real life between the lines of the story they tell.

Delusions share a great deal with ghost stories, too. 'Madame X' buttonholes her doctor with a story about a lightning strike down her back, and her utter conviction about the revelation she is dead. The story leaves us with a sense of gothic menace, pre-empting Freud's 1919 essay 'The Uncanny', or 'Unheimliche'. For Freud, delusions, like dreams, are stories from the unconscious. By the time 'Madame X' was at Vanves and sharing her belief about being dead with her doctor, spiritualism had become popular in Europe. Central to the religion was the belief the disembodied spirit could be contacted in the hereafter and seances were a means by which to bring the messages through. 'Madame X's worldview is Catholic, but her version of living death doesn't fit neatly into Catholic doctrine. She is even the conduit for an urgent message from the ether, in her case experienced as a jolt of electricity down her back almost as if taking part in a seance herself. She is passive, but she is also problematic, awkward and demanding further investigation.

Cotard saw 'Madame X's crisis as first and foremost a psychological one. We can't know what neurological element may have been in the mix, if any, but there were certainly difficult experiences in her past exerting an influence, though the stories come to us incomplete. She brings up a specific event in childhood but then keeps it out of view. She was an adult during the years of war with the Prussians.

Cotard is working before Freud's ideas arrive on the scene, but psychoanalysis would soon view trauma as the generator of madness, and uncovering the trauma, bringing murdered memories back into consciousness, as the road to cure. With her delusion she disconnects from this trauma. She removes these connections out one by one, disassembling her body so she doesn't have a stomach, or guts, so there's no need for any food, and she's got no soul either. She says she is damned but then that God and the devil don't exist either. She contradicts herself. She's made herself disappear, but then tells her doctor that she wants to burn alive, like a heretic. She is not as blank as she first appears, then, there's a faint request for redemption. But she won't wait. She passes judgement on herself.

Jules Cotard tried to cure her, but he ran out of time, too. He'd been at Vanves for fifteen years when one of the infectious diseases creeping through the population came for him and he died aged forty-nine on 19 August 1889 from diphtheria. He'd contracted the disease while nursing his beloved daughter at Vanves. As she endured the illness's characteristic fever, he had refused to leave her bedside for fifteen days.

Like Francis Spira three hundred years before her, 'Madame X' is reported to have starved to death.[13]

Portrait de femme voilée, Gaëtan Gaitan de
Clérambault, 1918–1919. Baryta print.

CHAPTER X

'Léa-Anna B' and the King: Grand Passions and 'Erotomania'

4 December 1920, Paris. An agitated woman in her early fifties disembarks a *Métro* train, still a relatively new system of transportation around the city. She approaches two gendarmes. She is being followed, she says. Other passengers are making fun of her. She demands protection. The police do nothing and the woman becomes increasingly irate and then lashes out physically at one of them. They take her into custody.

She is brought to the Infirmerie Spéciale, the psychiatric triage on the Île de la Cité, which is situated within a vast building housing a maze of public offices. It is here that a psychiatrist named Gaëtan Gatian de Clérambault will assess her through his favoured round-lens pince-nez spectacles, perched above a neat moustache. At forty-eight he's just a little younger than his interviewee. She's wearing a low-fitting cloche hat, pulled down like a helmet, a ubiquitous style which requires her to look out from under the narrow rim, giving her an independent, even defiant air. She is a milliner it turns out, so she knows her hats. The hemline on her dress is just high enough to show a flash of ankle, her top sufficiently low cut for the physician to notice. He asks her to explain her behaviour at the station. The King of England, she says gravely, is in love with her.

De Clérambault has just become the head physician at the Infirmerie connected to the Prefecture of Police. He's been physician there since 1905 so he knows its comings and goings. He takes his pen out of his breast pocket to jot down a few more details about her delusion. 'Léa-Anna B' will be her pseudonym.

She tells him about the moment she was certain of the king's attitude towards her. She now realised that he had tried to make his intentions clear through special officers who appeared disguised as sailors, or tourists, but she didn't understand until it was too late that they had been sent by the king. Looking back, it made sense that the knowing glances and the cryptic remarks were intended for her, but she had missed them. Many of her visits to Paris had coincided with high days and holidays, such as the fête de Jeanne d'Arc, celebrating the heroine of the Hundred Years War, and, of course, the fête de Noël. These festivities distracted her and she wasn't paying close attention. One day, when travelling by train, she met an officer from the retinue of General Lyautey, who revealed that he was an emissary from King George V. That was when she realised that the king was surveilling her, through secret representatives who had adopted a variety of disguises.

All the previous incidents suddenly fell into place. That knock on the door of her hotel room late one evening, for instance, must have been the king wanting a rendezvous. This misunderstanding had resulted in a delicate situation. Because she had not responded to his advances, he would naturally assume that she was rejecting him. Nothing could be further from the truth. What she needed to do now was to explain to him that she returned his love. So she set off for London to speak to the sovereign as soon as she could.

De Clérambault does a tough job. He oversees the well-populated area where policing meets medicine. Attached to the Prefecture's Infirmerie is a forensic psychiatry centre. He was chief psychiatrist for the Paris police through the 1920s and into the 1930s and he's a lawyer by training, so he plays a key role in describing, documenting and committing many of those who come to him through the system. He has been in post for fifteen years, conducting surveillance of those people who exist in the lowest social and economic strata: the high number of recent migrants to the city without a home or basic income; absinthe addicts, prostitutes, petty criminals. This is a good one. He asks her to continue.

The king began making advances in 1918, she said. From then on, she paid numerous visits to London so that she might speak to him and correct the record, spending extravagant sums of money on each trip. She waited around in the main railway stations imagining that the king would somehow get word to her of when they would meet. She prowled around the royal residences, finding her way eventually to Buckingham Palace and wandering around the perimeter there. She now believed not only that George V was in love with her, but that all of London, even the palace courtiers, knew of their affair and wanted it to succeed, including the blood princess Mary and her cousins who wanted her to become the king's mistress. She once saw a curtain move in one of the palace windows and interpreted this as a sign that the king was watching her. There were additional signs, but no direct contact.

Did 'Léa-Anna B's hunt for signs take her to St James's Palace, where Margaret Nicholson had loitered more than a century before, waiting for the usurper King George III? Certainly 'Léa-Anna B' shows a similar tendency to grandiosity as Margaret Nicholson did, in de Clérambault's version of events anyway. Nicholson did not believe King George III was in love with her – she was claiming her birthright – but both women express profound grievance at not being helped, and both are demanding undivided attention from the head of state.

When James Tilly Matthews received no response to his letters to the government, he turned up shouting during a debate in the House of Commons. 'Léa-Anna B' was ignored, on one trip after another to London. And, like Matthews, her thoughts began to darken. She came to believe that the king was deliberately scuppering her arrangements when she was in the city. On one visit to the capital, she says, the king had somehow organised it so that she would lose her way, and made her forget which hotel she was staying in so that she lost her bookings. He had also arranged things so that her trunk, which was stuffed with money and numerous portraits of the king, would go astray. The king had impoverished her so she had organised a sale of her furniture in the provinces. The sovereign foiled that, too, and had the auction frozen. She was still fearful about her money, and carried thousands of francs around with her on her person. Nonetheless, her passion for him is undimmed: 'The King might hate me, but he can never forget,' she says. 'I could never be indifferent to him, nor he to me… It is in vain that he hurts me… I was attracted to him from the depths of my heart.'

She continued spending money with abandon while she waited around for a word from the king, ran low on funds and eventually returned to Paris, angry, frustrated and with significantly less in her pockets. Then came the incident on the *Métro* train in Paris and she was frogmarched into the psychiatric system to meet the doctor who would record her case for posterity. She was experiencing a crisis of hope, de Clérambault surmised. He referred her to the psychiatric institution of Sainte-Anne, the same asylum in which 'Madame M' had received her initial consultation the previous year.

Her certificate of admission is brief and to the point: it mentions an alcoholic father and calls the family 'disunited'. There is no serious hereditary family illness, as far as the physician knows. But 'Léa-Anna B' is always lying, he says. She is an 'authoritarian', and her suspicions about the world have existed for a long time. She is 'proud', above all. Is this the same 'ruinous pride' that scuppered Margaret Nicholson? He calls her story of a royal suitor a 'paradoxical game'.

It is a game he is interested in playing. What does 'Léa-Anna B' want? He and his colleagues are at quite a pitch of writing up theories around delusions and presenting them to each other in lectures where they can argue out the points in public. From 1920, de Clérambault writes numerous articles about erotomania. He publishes nine case studies, with commentaries on a further three, in which he marshals his thoughts on a broader category of what he calls 'psychoses passionnelles'. He collaborates with Joseph Capgras on some of the cases, and they were well known for their sparring on the subject. The two are old associates. They both joined the conglomerate of psychiatric institutions known as 'Asiles de la Seine' in 1898 and continue to take turns as lecturer and audience for one another and to scrap about the interpretation of symptoms. It was de Clérambault who sat in on the demonstration of 'Madame M' and gave his tuppence worth on the cause of her delusion of substitute doubles. He suggested hallucinations were involved; 'Madame M' denied this in the strongest terms.

A few weeks after 'Léa-Anna B' is brought in, de Clérambault and another colleague, Brousseau, discuss her case at a session of the Société Clinique de Médecine Mentale and in 1921 de Clérambault publishes a landmark paper detailing the delusion.[1] He has been developing his thinking around erotomania for some time, and a few other examples have turned his head, but the case of 'Léa-Anna B' won't be put to bed. There is more he wants to talk about: the *modiste*, or milliner, is now fifty-three years old. She was well established in her work before she came to the hospital but she spends a lot of time completely 'idle', having apparently lived an easy life thanks to the grace of a rich and high-ranking lover.

Where did she come from?

'Léa-Anna B' was born in the late 1860s, not long before the outbreak of the Franco-Prussian War.

The years 'Léa-Anna B' made hats coincided with the heyday of the millinery trade in Paris at the end of the nineteenth century. This was an industry which employed thousands in small, independent millinery shops, often as not run by women. If you were in that line of work you often had jobs in dress shops, too, in the hat departments of Galeries Lafayette, for example, or Le Bon Marché, the art nouveau temples to retail which offered a touch of the sparkle of couture at more affordable prices. 'Madame M' was just a couple of years younger than 'Léa-Anna B' and also making her way in the city at this time. They could conceivably have crossed paths somewhere on the teeming boulevards where dresses and hats were made and sold. 'Léa-Anna B' was well established in her métier by her early twenties. Working women like 'Léa-Anna B' didn't usually have much of an education before they started an apprenticeship and, after years of hard work that didn't promise to lift them out of their circumstances, many looked for financial security in a relationship, a perfectly legitimate route.

When she was twenty-two 'Léa-Anna B's circumstances changed. She became the mistress of a rich and well-connected lover. The relationship lasted eighteen years. When her lover died in 1907, she enjoyed her freedom, as well as using it for what de Clérambault calls 'personal benefit'. She quickly got together with a much younger man. Her doctor suggests he was easy to dominate. He also, apparently, owned a castle.

He bought her a house a fair way from Paris, and they spent the war years there together. She enjoyed some hospitality in local society as a 'kept' woman, but she became lonely, suffering from being uprooted from the city and transplanted into the sticks.

Given how badly adjusted she was to rural life, it's a fair bet that 'Léa-Anna B' wasn't originally from the countryside. People who migrated into Paris from provincial towns typically didn't have money to fall back on. She had become a rich woman, but she'd been a mistress for eighteen years rather than a wife and although seemingly happy enough with the arrangement, it would have meant living with a degree of insecurity. The choice of a new lover, after the first died, represented even greater heights in terms of status and money. She fully intended to marry this man, she tells her doctor. According to de Clérambault, 'Léa-Anna B' spent too much time on her own. She makes the very mistake

Robert Burton warned against in the *Anatomy*. Without enough to do, she begins to feel alienated from the world she knows.

This was the period when 'Léa-Anna B's trips to London began in earnest, coinciding it seems with a developing love affair with England and its different customs and diversions. She would stay in luxury hotels and invite numerous godchildren to town to show them a good time in the restaurants, cinemas and department stores. Each time, however, she had to come back to her isolation in the French countryside.

At forty-three she began to complain of organised persecution by the local farmers, who were accusing her of sexually corrupt-ing this young man she lived with. The new relationship brought money, but it also brought mounting moral censorship from the community. And he did not marry her, despite her express desire to become his wife.

She'd apparently shown symptoms of paranoia before the war, but 'Léa-Anna B' described constant persecution from 1913 onwards, as well as fits of jealousy. She believed that she was being followed and spied on. The farmers played pranks on her, harassed her, sent thugs to insult her, followed her everywhere. Once the war had started, when she was forty-eight and believing she had been denounced as a spy, she destroyed foreign govern-ment documents during a fit of 'spite' that lasted for six weeks, she said. In 1917 she came to believe that an American general who commanded a nearby army base during the war was in love with her. On many of her extravagant trips away from home, she thought she was the object of silent advances by officers of many ranks, and regretted not having taken advantage of the situation. She offers a long list of the other things her neighbours said about her, their campaign to sully her reputation escalating. The paranoid and grandiose thoughts intensify.

'Madame M' reconciles her personal losses with a world in which her loved ones have been replaced by doubles. 'Léa-Anna B' creates an alternative reality in which she is a critical component in the war effort, the keeper of military secrets that other people want and she must burn, in the manner of a fictional heroine, in order to save lives. It is her mission to expose the conspiracy of the farmers to blacken her good name. She is at the centre of things, and she's an object of fascination and attention, even if it is of the negative variety. She has found the perfect setting in London, another glamorous city, not too far away, with a state-of-the-art underground train system much like the one in Paris. London, though, is a new society – she's found herself on the wrong side in France. The city allows her to reinvent herself.

And then in 1911 her relationship with her lover ruptures after four years together and she unceremoniously, publicly, loses her only social defence against the neighbours.

'Léa-Anna B's fortunes changed again in her own eyes at least, this time reversing in the wrong direction, from high to low status. She was rejected, humiliated. It was the breakup, her doctor speculated, that set off the full-blown delusion,

More probing from de Clérambault leads to a few key discoveries. One of the people who publicly denounced her as a spy in the countryside was her priest. There was also an incident when she stood accused of being underdressed in public. A low décolletage had been necessary, she said in her own defence, since the onset of her angina which followed diphtheria. She undresses to relieve a rapid heartbeat. She also mentions having had jaundice for six months in 1915.

She has come to believe there are machinations against her, instigated by what she refers to as an 'ancient family' in the country. De Clérambault wonders if it's the family of her ex-lover. She only gets this hostility in Paris, not London, which is another reason for the allure of the English capital over the French. A sense of eroticism and exhibitionism weaves itself in and out of her account. We sense this sexuality has also been the root of shame.

Then, into this world steps another high-ranking figure. They don't come any higher, in mortal form anyway, than the King of England, and he is in love with her. Her delusional beliefs build to an erotic crescendo starring George V, which is only accelerated by the deafening – she would call it pregnant – silence from the palace. 'Léa-Anna B' performs some serious contortions of logic in order to sustain her belief in the king's love, given that he's been ignoring her. The king *was* in love with her, she explains, but since he's now rejecting her, it must be hate rather than indifference; the kind of hate that is an extension of grand passion, and the feelings she ventriloquises are as strong as they ever were. There is yet more grandiosity. The king is just one of a long line of men who have been in love with her: officers of many ranks, the American general, another king even, this time the Belgian monarch, who wrote letters to her. The nature of the rejection from her point of view is also interesting: she is ignored only because she didn't see the signs or respond to the men's advances. In other words, *she* rejected all of them first.

It was Hippocrates, the 'Father of Medicine' who first described the kind of 'raving love' that Galen attributed to an imbalance of the humours. Characters whose senses are wildly disturbed by love, like Dido prostrate with love for Aeneas, or Helena pining for Demetrius in Shakespeare's *A Midsummer Night's Dream* have a place in the collective romantic imagination. Shakespeare's play dramatises the strange subjectivity of love, which is rarely conveniently returned. A fairy potion dropped into an eye can alter perception and make a person fall in and out of love at first sight. If only it were that simple.

The idea of 'love melancholy' as a kind of mental illness first appears in literature in 1610 in a work by a physician from Toulouse, Jacques Ferrand: *Erotomania, or a Treatise Discoursing of the Essence, Causes, Symptomes, Prognosticks, and Cure of Love, or Erotique Melancholy*. It describes the physical manifestations of the 'unreasonable' tides of happiness and sadness that accompany love, and recommends treatments to calm this passion. The Inquisition was enraged by the contents of Ferrand's book, principally because from their perspective lovesickness was in the soul, and therefore a religious matter which Ferrand had no business meddling in with his earthly cures like changes to diet and exercise regimens. They also objected to Ferrand's discussion of sex, as well as mentions of astrology and palmistry as a way of reading the future, even though he carefully debunked these claims as he went along. Robert Burton possessed a copy of the 1623 edition of Ferrand's work and seems to have been profoundly influenced by it: a substantial chapter of the third part of Burton's *Anatomy* is entirely devoted to 'Symptoms or signs of Love-Melancholy'.

Burton describes the mental disturbance brought about by infatuation, calling it 'a plague, a torture, a hell, a bitter-sweet passion at last' and, fundamentally, a 'perturbation of the senses' with clearly identifiable symptoms. He includes in the *Anatomy* most of the world's great love stories by way of example, from the ancient myths, to contemporaneous plays and poems. Love has been the realm of poets and playwrights, but both he and Ferrand are looking for the pathology. 'Love Melancholy' may produce sonnets in some, and can be a sign of a refined soul, but it can also be an illness which requires treatment. Burton uses up a good deal of ink and page to warn of the dangers to mental health of sexual frustration in a woman. This idea is still running four hundred years later when, in 1921, the British psychiatrist Bernard Hart refers to erotomania as 'Old Maid's Insanity'.[2]

*

An understanding that the mind can be profoundly disturbed by infatuation has existed as long as there have been people in love. From early childhood we are presented with a pantheon of literary figures tortured in one way or another by unrequited love. What 'Léa-Anna B' presents us with, however, is something different. 'Erotomania' as described in the early twentieth century was the belief that someone else was in love with *them*. It's a twist on the old trope of a person in love to an excessive degree, usually with someone who doesn't feel the same way. The nineteenth-century madness of unrequited love, the obsessional love or 'monomania' that Jean-Étienne Esquirol had written about in 1845, was rerouted and tuned back on itself. Joseph Capgras and Paul Sérieux, in their joint work on disorders of interpretation of 1909, saw this as another example of a person reading the world wrong, fixing on someone they have never met but who they feel they know because of the mysterious clues they are picking up which they, and only they, can interpret. In 1913, the German psychiatrist Emil Kraepelin talked about the delusion of someone else being in love with you as a kind of paranoia.

'Léa-Anna B' also told de Clérambault about an extraordinary system called 'La Morve' which gives form to the conspiracy against her. It translates literally as 'snot' and this strange material moves around crowds affecting behaviour. 'La Morve' was how she explained the numerous exchanges of signs between strangers that she said she witnessed. Like the magnetic rays from James Tilly Matthews's 'Air Loom', it's an organising concept for a secret plot, but 'La Morve' manifests something like the clouds of ectoplasm that mediums produced at seances. The people who follow her make 'sniffling noises', and she names them 'La Morve', too. She cannot escape, she says, and her situation is a 'flying prison, mobile cage, and perpetual chains'. 'La Morve' tracks her and envelops her, expressing her tightening sense of claustrophobia.

People often describe something like a ghostly apparition when recalling how their delusion first appeared. A person bears witness to a strange reality but the people around them don't believe a word of it. It is wildly improbable, but they are unshakably convinced of its reality and the urgency of the revelation even if they can't convince anyone else of its authenticity. Disproving evidence won't shake a delusional belief any more than turning on the light and checking behind the curtains will make belief in a phantom evaporate.

The history of ghost sightings throws up time-honoured stock themes and 'types': the unquiet ancestor shrouded in white

avenging an unresolved injustice from the past; the petulant pol-
tergeist. Delusions, similarly, offer us familiar motifs which recur
again and again over millennia. The beliefs are easily debunked,
but the stories are about more than that. Just as there is social
history in ghost hunting which is not contingent on whether ghosts
can be proven or disproven, certain themes acquire meaning
from the hundreds of years people have repeated them in their
delusions and passed them down the generations.

'Léa-Anna B's life before she meets de Clérambault is sketchy
and difficult to make out. But there is an event she relates to her
doctor which stands out, like 'Madame X's first communion, and
'Madame M' catching sight of the uniforms being loaded onto
the truck or the unusual appearance of the baby's fingernails
suggesting poisoning. 'Léa-Anna B' mentions it only in passing
and it's separated from its context, but it interests her doctor. She
had, she said, burned a collection of prints, of her relatives in
the first Boer war. She told him she had looked after the prints
with care, until the date of the burning. De Clérambault fails to
join up this event with the other things he knows about her, and
this burning remains a cryptic act, with a backstory we cannot
access. De Clérambault has called her a habitual liar, but it is
hard to see why she would lie about such an event and give such
strange and specific details when there's no obvious reason to do
so. She doesn't try to explain it or use it as justification for her
behaviour. If we can take her at her word – in this instance at
least – it appears she did have relatives who were involved in the
Boer War of 1880–81. The fact that she had carefully kept their
pictures for at least twenty years, and then dramatically burned
them, suggests that de Clérambault's note about a 'disunited'

family was on the money, and that her feelings towards this family were at the very least complex and unresolved. The questions go unanswered, but leave us with a haunting imprint of deep, dark family secrets.

There are further hints of skeletons rattling in her family's cupboard. When de Clérambault conducted one of the early interviews with 'Léa-Anna B' he asked her brother-in-law and twenty-eight-year-old niece to act as witnesses to help test her lies. It's a very uncomfortable setup and the skeletons begin to rattle as she explains to these family members why she's ended up in trouble with the authorities. The doctor reports that she immediately starts spinning them a yarn. She says she has simply been taken ill with something or other, and brought in for a cure. She is all smiles, all affection. De Clérambault said the team would have been fooled by her if they hadn't known better. She writes down a statement for the doctors demonstrating very poor writing and spelling, but she knows what she wants to say, and the document confirms to the doctors that she is free from 'graphorrée' (the compulsive scribblings of a 'mad person' which they are trained to look out for). In return for her freedom, she pledges not to go down into the *Métro* again and to dress more modestly from now on. She reproaches the doctor for noting down that her father was an alcoholic. He was wrong to do this, when she had just opened 'the garden of her heart' to him and his colleagues, and he's a hypocrite for shaking her hand. She reassures him that they really don't need to worry about any future bad behaviour. She wants to live in England because she prefers the laws there. She is an Anglophile. It's as simple as that. She will be travelling there with the Thomas Cook travel agency, and if she does go to Buckingham Palace she'll be discreet, she promises. She has plenty of pictures of the king in her case to be getting on with. Again, says her doctor, here she is lying with a natural ease. She accuses him of abusing a pretty Parisian milliner. He scoffs at her efforts to present herself as chaste. He doesn't believe a word of it.

There is more to de Clérambault than meets the eye, too. He was born just a year after the Siege of Paris, and moved to the city at the age of thirteen to study at the Collège Stanislas. This was an interesting moment to be arriving at a prestigious private Catholic school. The Jules Ferry Laws had just taken control of education away from clerics and the college was a dinosaur that would have to fight to survive. Like 'Madame M' and 'Léa-Anna B', he spent his formative years after the Collège Stanislas in a city bursting with creative energy. This perhaps goes some way to explaining his parallel passion for, and expertise in, the visual arts. Alongside his day job he lectures on the art of draped costumes at the prestigious École des Beaux-Arts in Paris and promotes the idealised forms of ancient Greece and Rome.

And he has secrets, too. If you were to go to his home, unbeknownst even to his closest friends you would find lengths of fur, silk, velvet, satin, taffeta and even a mannequin swagged in these fabrics. Given his study of draped costume it's safe to say that he knows his way around the Paris ateliers, and he knows about millinery and the styles of the day. Let's not forget he admitted a couturière at the Infirmerie back in 1918 who was ranting during the interview about how her husband had been replaced by doubles.

He's a good liar, too. By way of an attempt to cure her of her delusion, de Clérambault informs 'Léa-Anna B' that she is to

appear before a committee of eminent men. Their reputations
extend all the way to England, he tells her, and he encourages her
to take this opportunity and plead her cause in front of them. He
stages a mock demonstration, pretending that he will be able to
arrange a meeting with the king, although he says he's not sure
it's a good idea. He wants to see if she will be able to control
herself were she ever to come face to face with her admirer. The
demonstration is reported in direct speech like a play in *Erotomania*:

> 'I'll keep my hands like this, behind my back, so you can stand
> behind me and restrain me.'
>
> 'I'm also afraid that you'll immediately throw your arms
> around his neck.'
>
> 'But you can hold me back.'
>
> 'Yes, but what will the princesses say about all this?'
>
> 'They probably won't even be there.'
>
> 'Didn't you tell me they were very interested in all this?'
>
> 'That is something between him and me.'[3]

At de Clérambault's suggestion, 'Léa-Anna B' retires to write a
letter to the king and after a quarter of an hour she confidently
hands it in. In it, she opens her heart to the king, humbly assuring
him of her deepest affection and that she reciprocates his feelings
for her. She tells him that she's hoping he will arrange for her to
come to England. It is signed 'L. Anna B à l'hôpital Ste-Anne.
Paris, le 20 Décembre 1920'. De Clérambault tells his colleagues
that a ruse such as this invariably works with a patient like 'Léa-
Anna B'. The old ones are the best. We raise an eyebrow at his
confidence. This belief in ruses as legitimate, long-lasting cures

will look different later in the century after landmark cases like the
Three Christs of Ypsilanti and the ethical mess they left behind.

Capgras wrote up additional cases of people who believed
others were in love with them when they were not. The notes
are brief and fragmentary, we catch glimpses of eyes and ankles
only, and they offer very little context, but nonetheless some clear
themes emerge.

Capgras describes the 1921 case of 'Léontine D', a twenty-
eight-year-old worker who believed that a captain wanted to marry
her. Another woman, thirty-three-year-old 'Renée Pétronille S',
insisted that a government clerk was in love with her, kept watch
over her and was having her followed by prostitutes as well as
some subordinates under his instructions, over a period of seven
years. Apparently, this began when she was turned down for a
'safe conduct' pass to travel during the war in 1915, and then,
after some flirtatious approaches, she became violent, and the
letters started. She didn't deny that he was married, but suggested
they live in a *ménage à trois*.

The love-struck authority figure in one case is a doctor, and in
another a priest. 'Clémentine D' had neighbours who were trying
to control her through electromagnetic machines. De Clérambault
also describes the particularly striking case of 'Henrietta H' from
1923. Over a thirty-seven-year period, this fashion designer, now
fifty-five, had the recurrent belief that a priest both loved and
persecuted her. The erotic delusion began suddenly during mass,
and from then on she harassed the priest remorselessly. The family
tried to deal with this by marrying her off, but less than a year later
she began a series of affairs. Tranquil periods were followed by
reawakenings of her love, when she would return impulsively to

Paris to pursue this same priest again. After a divorce, she worked as a domestic servant, but only in households with a telephone, and continued her ambushes, scenes, letters and calls.[4] There is one instance of a man suffering from erotomania among de Clérambault's notes. In 1921, 'Louis G', a thirty-four-year-old 'fitter', denied the legality of a divorce from his wife, and came to believe that she was pursuing *him*.

Patterns reveal themselves: the women in the notes are more often than not in menial jobs: in factories, working piecemeal as seamstresses or in service, while the 'enamoured' males are medics and clergy.

De Clérambault's observations on 'Léa-Anna B' – the first clinical account of erotomania – coincides with the birth of Hollywood and its spotlight on romantic love. Vulgar melodramas playing out stories of grand passion and all sorts of variations on the theme were up on the silver screen, larger than life, and for the first time people were living in their reflected light. Hollywood had distributors around the world and dominated the international market, which the French had led before the war with their more experimental feature filmmaking. The motifs of true love from Hollywood's golden era arrived in cities and towns. The idea that romance was linked with destiny – that two people were meant to be together – was put in front of people who had until now been more pragmatic about such matters, at least in public, and certainly more modest.

A woman like 'Léa-Anna B' would have had plenty of opportunity to see some of the most popular films of 1920: *Romance* was a silent drama starring Doris Keane. The film featured a priest (yet another) who was in love with an opera singer, and centred on the conflict about whether he should rise above worldly matters or leave with the woman he loves. *The Restless Sex* featured Stephanie, an appropriately restless and adventurous young woman, who is involved in a torrid love triangle. There is something quintessentially twentieth century, then, about this delusion. 'Léa-Anna B' creates a Hollywood fairy tale for herself to star in.

Outside the cinema, things were very different. In 1920, London, the place that for so long had been 'Léa-Anna B's entertaining, diverting, escapist world of cinemas and department stores, was a city in shadow.

On her last excursion to London in the winter of that year, on her way to Buckingham Palace, 'Léa-Anna B' would have found herself navigating a city trying to come to terms with the staggering losses of the war. If she visited the city on one winter morning in particular, 11 November 1920, she would have been swept up by one of the most remarkable events of the century: the funeral of the 'Unknown Warrior'. It's plausible (though in the balance of probability it's unlikely) she was there that particular day, but whichever day she disembarked at Dover that winter and took a train into Victoria station she found a population mourning its dead. That particular day, 11 November, the mourning had a focus. A soldier had been selected from several unidentified bodies exhumed in northern France and Belgium and brought from Arras. Like her, the body had travelled to England by boat, via Boulogne, and then on a train from Dover, to be laid to rest in Westminster Abbey.

If she did choose the 11th for her trip, she would have been able to watch the vast, solemn procession pass, and perhaps glimpsed the coffin on a bier, wrapped in a flag, and King

George V himself following on foot, all the way from the new Cenotaph to the Abbey. The nameless soldier is to be buried among kings. She would have seen the hundreds of people who had gathered before dawn, standing in the light of the gas lamps lit early for them, outside the northern door of the Abbey before it opened at eight o'clock; queues of women and children, grieving their own sons, husbands and fathers, extending beyond Victoria Tower. More than a million would queue to see the tomb in the weeks and months that followed and even if her last trip to London that winter was on another day, she could still have seen the queue, snaking back from the Abbey door for a hundred yards. She could well have shared her hotel with some of the mourners. The image of a king, humbly on foot, following an anonymous soldier who is to be buried with royalty, would have struck any bystander as a powerful role reversal, which raised up ordinary lives and their extraordinary sacrifices. It's a tableau that speaks wordlessly of the disruption that the war precipitated; a war that cut down a generation of men and cut millions of romances short.

De Clérambault and his patient were both changed by the war. She'd been isolated in the countryside, but felt the war close by. She met the officers of many ranks and the American general and developed delusions that these officers were in love with her as well as a belief the locals were after her. Already an acclaimed painter and photographer and a pioneer of psychiatry, de Clérambault had fought in the war, serving with the artillery even though he was supposedly three centimetres too short to enlist. At his own request he was sent to the front, and was seriously wounded on two occasions. He was awarded the cross of the *Légion d'honneur* for his exceptional bravery.

When they came together after the war, most conversations were private, one-to-one negotiations between the doctor trying to understand and describe the delusion in a clinical sense, and his patient, describing it from first-hand experience. The power balance could shift back and forth in a few words. In this case it's hard to escape a hint of eroticism in how this doctor considered his patient. He wants to look; she wants to be looked at. De Clérambault had seen action on the front line, like so many of his peers. His patient had lived through the same times, the same war, from a different vantage point and a contrasting social standing.

During the war de Clérambault volunteered for a tour of duty in Morocco, which was then a French protectorate, and he returned to the country after the war to photograph Moroccan robes, capturing more than four thousand images, which he used in his lectures on drapery at the École des Beaux-Arts in the 1920s. These photographs show faceless apparitions, their form created only by the folds of the shroud. You feel as though you are seeing something which is not there, catching the shape of a person only, their shadow, as you might be tricked by the folds of the bedroom curtains. They are ominous, reminiscent of the unquiet spirits who appear on the galleries of country houses to avenge a past wrong.

De Clérambault's psychiatric specialism, experience in police work, and his interest in the properties of material all converged in a niche investigation. He worked for years on the psychology of silk fetish – specifically women who stole lengths of silk from department stores and took them home to masturbate with. 'Kleptomania' was seen as a particularly female activity, stealing for thrills, but also as a way of being seen. In other words, in the hope, maybe unconscious, of being caught in the act.

After his death, friends clearing de Clérambault's house discovered his own private collection of fabrics, in all manner of tactile and luxuriant materials, as well as the mannequins he dressed up in the cloths, strongly suggesting he had crossed the line from observing the fetish in others to indulging in it himself.

He practised the martial art of jujitsu and was also renowned for a quick temper, on one occasion demanding a duel when he believed someone had stolen his ideas. With bitter irony, given his passion for looking at the world and its representation, he was plagued in late middle age with failing eyesight caused by cataracts.

On 17 November 1934 he shot himself, in front of the mirror, with his service revolver. It was as if he had contrived the scene of his death to be a piece of performance art. He dramatised the way different realities are created depending on who's looking. Other doctors gave 'Léa-Anna B's delusion the name 'De Clérambault syndrome' after his death.

Did 'Léa-Anna B' get what she wanted? 'De Clérambault syndrome' is in one sense, clearly, a wish fulfilment. It rearranges reality to boost a crushed psyche. De Clérambault came to see it as an answer to the demands of sexual pride that life had failed to provide. Emil Kraepelin picked up this melancholy refrain and called it 'psychological compensation for the disappointments of life'. It provides an explanation for an uncomfortable cognitive dissonance: a conflict between where a person feels they are, or should be in their personal relations, and where they actually are. 'De Clérambault syndrome', like a delusion of grandeur, is also a power grab: the adoration is 'given' without the permission or reciprocation of the person who is supposedly feeling it. 'Léa-Anna B's claims about the King of England are preposterous, but she is in control, putting thoughts into someone else's head, and free of the emotional risk, responsibility or regret that comes with being in love, saying so and having to wait for a response.

Psychoanalysis has tended to interpret the belief as a strategy to protect against the distressing sense of being unloved. Freud also described it as a defence mechanism against unwanted sexual impulses which lead to the 'projection' of feelings onto others.

It's now understood that both men and women experience the delusion. Men with 'De Clérambault syndrome' are more likely to act in a way which brings them into the criminal justice system – it is now called stalking – and so fewer men relative to women come in front of psychiatrists to be counted. It's true, too, that women will sometimes have been labelled 'deluded' where men might not have been. The concept of 'hysteria' was a broad catch-all diagnosis exclusive to woman which connected biology to an innate instability and irrationality in matters of love, passion and sex. Cases of erotomania would easily have been caught in its net and by definition a man would not have been in their number.

'Léa-Anna B's is a quintessentially twentieth-century delusion. Individuals in significant numbers began to confide in physicians about the rich and powerful strangers who were in love with them, and many would be forced explain their conduct to the police. This delusion, like paranoia, which is the most common delusion seen by doctors today, must be taken seriously. Being unshakably, and wrongly, convinced of someone's love, or their villainy, is a serious threat because it justifies future action while denying any responsibility for it. It is sobering to look back at 'Léa-Anna B' from the vantage point of a century's distance when 'celebrity stalking' is a pernicious problem, with male as well as female perpetrators. Reality television shows and social media have further blurred the boundaries between real and imagined relationships and encouraged delusional thinking. 'Léa-Anna B's delusions, both paranoia and erotomania, went on to define an age we are still living in.

De Clérambault does not report what happened to 'Léa-Anna B' after she wrote to the king. He suggests that his ruse was successful. Writing a letter gave her some agency, recorded and witnessed her beliefs, and allowed her mental stability to return. It is possible that she was released from Sainte-Anne at some point, but the records do not continue. She got some of the attention she craved as small compensation for life's disappointment and hurt pride, so the delusion succeeded on limited terms, for as long as we can follow the story anyway.

As with so many other cases, the lesson here seems to be not to try to explain away a delusion as madness, but to treat it as something important and compelling, even marvellous, like a glass vase or a cloud of vapour or gas that swirls around influencing people. When someone with a delusion is offered kind attention, even the most strident delusion will lose some of its form, dissipate like a mirage, a trick of the light, or vanish into thin air like a will-o'-the-wisp.

Conclusion

Many people with delusions, especially if they are poor, will see police officers a lot more often than they'll see a sympathetic doctor. That was true in Paris and London in the eighteenth and nineteenth centuries and it's still true.

Most of our subjects found doctors who wanted to listen and bothered to record the conversations. They leave us a rich legacy. But after following these life stories from the past, which are inevitably incomplete, we lose the trail and the first question is: did the delusions continue? We do not discover for certain if a person was cured long term, or what sort of life they had after their time in hospital, when they set off back down those dimly lit streets. We are left with snapshots from their lives, out of context, and the odd scene stands out, crystal clear and memorable in an otherwise fragmented and impenetrable life story. We pick up touching personal traces as we go: signatures and marginal notes in the admission ledgers and birth and marriage certificates. We see a name on the page in a particular hand, wonder if it's really their signature, and feel sad when we spot a matching name has 'moved to the incurable' part of the hospital, or a sense of joy on the rare occasion we see a release date. Each of these case studies has gaps in the story, and we can't know what's really going on in the sufferers' minds or what undiagnosed biological factors there may have been. Attending to the words and images in the delusion itself offers the best chance of detecting a true-life story.

Many of the physicians from the past did understand, in one way or another, that if they pulled up a chair and listened as their patient told their story, it did seem to help. Sitting with someone who is experiencing a delusion, venturing some of the way into their world, even if it's through a ruse, can forge a connection. They may not come back from that alternative world, but they will share, confess. Even ruses or 'white lies' could succeed, for a time, as with 'Léa-Anna B' and her letter to the king. There's an essential dignity in what each of them is asking for – that we consider their most profound needs: to be worthy of attention and interpretation and to be reconciled with their circumstances. The lives of these individuals span six hundred years, but they have a lot in common. They all want to find some peace, respectability, love. Standing next to each other in this book they each have something to say to each other as whispers of solidarity pass between them, and to us.

Delusions are very articulate in their own right. 'Madame M' turns the loss of her children and her sense of alienation into substitute doubles who stalk her around Paris looking just like her and her missing children. Another woman's total bewilderment is communicated with consummate eloquence in a letter in the persona of 'Madame Bonaparte'. She invites her husband 'Napoleon' into her inner life where magnificence, real suffering, intimacy, alienation and all sorts of other contradictory sensations co-exist. She can't find the words, she says, and yet she does find the words, dressed up in her 'delusion of grandeur'. Clothes become important in all of these stories – hats, uniforms – because delusions are all about telling people how to view you and how to treat you. In many ways, delusions are a game of dressing up.

With delusions, the messages themselves are clear. They give form to dizzying ambiguity and confusion, whether that form is a king made of glass, or the proud family tree of a disgraced servant displaying multiple scions of European royalty, or a body with no guts or brain. There's artistry and dignity in the endeavour, and dignity is not something that people experiencing delusions have traditionally been allowed much of.

What's behind them? At a psychological level, low self-esteem, excessive worry, trauma, and, in the case of paranoia, isolation are contributory factors. A feeling of belonging vaccinates against delusions.

Other themes have also stood the test of time: money, love, war, uncertainty, shame, social anxiety, technological progress and, above all, inner conflicts, be they personal or religious. Delusions offer escape from a wretched situation; they repackage wounded pride, humiliation, or shame at the hands of lovers or neighbours.

Delusions sit at the intersection between neurology, and psychology, and different causes frequently co-exist within the same individual, overlap and fuse. We can see that many of these people in the historic cases were experiencing several different 'types' of delusion at the same time. Minds are not neat, and, thanks to neuroscience, we now know that, with many delusions, organic brain disease exists alongside psychological causes.

Each new frame for delusion in the centuries that follow tells us something important about man's place in the world at that time. In the literature of the classical world, the gods had the upper hand over mortals, controlling even their delusional follies by proxy. Then the Christian God took charge of the Western world and religious authorities quickly decreed a link between delusion and demonic possession. From then, right up until the nineteenth century, it was religion that decided the fate of the people experiencing delusions. Since then, diverse medical and psychological disciplines have had a go at interpretating the phenomenon, argued over it, filled in some of the blanks. Psychoanalysis prized them as creations born out of childhood experiences that had lodged in your unconscious mind. Biological psychiatry in the 1960s and 1970s downgraded delusions again to evidence of how a diseased brain jumbles things up and, as it were, spits them out. We haven't solved it yet.

We all have at least one belief about the world or ourselves that the others who know us well would not agree with. We are all somewhere on the scale, most, of course, not as far up it as the individuals featured in this book.

It's clear, though, that many delusions are helpful to us, can even be critical to our psychological survival. Again and again, delusions organise the enemy and give us a very clear purpose which might be, for example, to expose a treacherous conspiracy, or get the king to admit that he has stolen the throne that is rightfully ours, or demonstrate that despite our sense of alienation we have in fact been loved passionately for years.

People cling to delusions as a life raft. Be they Charles VI of France in the fourteenth century casting himself apart from the crowd wrapped in blankets, or Robert Burton a little later on navigating the dark sea of his life with reference to the stars of his horoscope. You can't therefore reason someone out of a delusion by explaining why it's not true. It's their best chance of survival.

At this particular moment in history, at the beginning of the third decade of the twenty-first century, paranoid delusions in the form of conspiracy theories have taken an even firmer hold on sections of the global population, and they, too, simplify a complicated world. Uncertainty faces most of us at points in our lives, and always has, but some people are better able to accommodate ambiguity than others. Some just aren't equipped to tolerate it.

The conversations between the subject and the doctor in this book spark with symmetries and sympathies. The physicians are attempting to understand and cure their patients and at the same time trying to find their own equilibrium, their own way through, many having hard, first-hand experience of war. On the surface they couldn't be more different. The points of contact are touching and, sometimes, every bit as intriguing as the delusions themselves.

Delusions are creations of the imagination, and they demand attention much like pieces of performance art, or raucous protests. They are forged out of inner conflict and they do a lot of work for us, organising and articulating perfectly ordinary wishes and fears, expressions of anxiety and suffering. Above all they are protection from an unbearable reality; not so much a ticket out of town as a regeneration and rebranding of the old one. They give shape to a chaotic inner life. They allow us to disconnect and check out. They puzzle and haunt us. They are at one and the same time completely understandable, and utterly mysterious.

Acknowledgements

I'm indebted to the clinicians whose work I've read and who I've spoken to over the course of writing this book who have helped me orientate within the many different disciplines involved in this unwieldy subject over the centuries.

I owe particular debt to Professor Daniel Freeman, clinical psychologist at the University of Oxford, who is pushing an ever more nuanced understanding of brain and mind and with whom I made a series in 2018 for BBC Radio 4. He led a series of conversations for the programmes with people who generously shared their personal experience of delusions. I admired his skill and empathy and what I learned from these contemporary conversations was invaluable in helping me notice patterns, and understand more about the people in this book who experienced delusions a long time ago in the past. Heartfelt thanks also to all the people who told their stories.

Eve Streeter produced the BBC Radio 4 documentary series with me. A great friend and I've learned so much from her great skill and sensitivity over many years working together making radio documentaries.

My interest in delusions really began when I spoke to Andy Lameijn in Leiden about glass delusion. I'm grateful for his generosity in sharing his experiences with me, not to mention his time, and for giving me the three antique volumes of Robert Burton's *The Anatomy of Melancholy*, which I treasure and dip back into for inspiration daily. It's an intoxicating subject and it's been marvellous and sometimes head-spinning to explore – I've been happily lost in each of these worlds, with ten new companions to show me around who have intrigued and entertained me, sometimes led me astray, and frequently moved me. They feel like friends now.

Mary Ann Lund's work on Robert Burton and melancholy in the early modern period is second to none and has been an invaluable resource and inspiration both for the radio series and the book. My conversations with Adam Phillips about glass delusion fired my interest and guided my thinking. I owe a great deal to Laure Murat for her painstaking research which unearthed so much from the archives in the making of her political history of madness.

Mike Jay and his body of work got me hooked on James Tilly Matthews and the Air Loom and *The Influencing Machine* which sets the bar for scholarship of Matthews.

I'm also grateful for my conversations with Andrew Scull around the history of madness and delusion, about which no one knows more.

Andrew Hussey's writing about Paris gave me a passion for the darker side of the city's streets and made me want to return to them. Hallie Rubenhold's *The Five* introduced me to forgotten lives in Victorian London. Hilary Mantel's novel *A Place of Greater*

Safety transported me to the French Revolution and introduced me to people I hadn't met before and now feel as if I know. Horatio Clare's riveting account of his own experience and his investigations into treatments, *Heavy Light*, was a game changer.

I'm enormously grateful to Emily Lutyens and Kate McCrickard for painstaking research in the Paris archives while the city was in lockdown, for batting the subject to and fro over the Channel, and for the extraordinary articles, photographs and postcards they found for the book.

Richard Bentall's research on paranoia has been both fascinating and really helpful and Professor Edward Shorter, a historian of psychiatry from the University of Toronto, gave me an overview of the history of psychiatry, and specifically catatonia. Thanks are due to friends and colleagues for lively discussions along the way; to Joby Waldman and others at Somethin' Else, and BBC Radio 4 for support getting the documentaries on air.

I owe a great deal to my agent Luigi Bonomi for championing the book in the first place and for all his encouragement.

I'm so very grateful to Richard Collins and to Polly Hatfield and Rida Vaquas at Oneworld for their careful and clever readings of the text and everyone who has had a hand in getting the book together and into such beautiful shape: a dream team.

At home, Mohit and my parents and wider family, Jo Spence, Nora Rose and Esmee, thank you for entertaining the little one while I wandered off into other centuries with these intriguing characters.

Above all to Sam Carter, my brilliant editor at Oneworld. His support, guidance and enthusiasm for the subject has meant everything.

Select Bibliography

Andrews, Jonathan, 'Bedlam Revisited: A History of Bethlem Hospital, *c*.1634–*c*.1770' (Ph.D. thesis, London: Queen Mary and Westfield College, London University, 1991).

Andrews, Jonathan, Briggs, Asa, Porter, Roy, et al., *The History of Bethlem* (London and New York: Routledge, 1997).

Andrews, Jonathan, and Scull, Andrew, *Undertaking of the Mind, John Monro and Mad-Doctoring in Eighteenth-Century England* (Oakland, CA: California University Press, 2001).

Appignanesi, Lisa, *Mad, Bad and Sad: A History of Women and the Mind Doctors from 1800 to the Present* (London: Virago, 2007).

Berrios, G. E., and Kennedy, N., 'Erotomania a Conceptual History', *History of Psychiatry*, 13(2), 2002, pp. 381–400.

Cahalan, Susannah, *Brain on Fire: My Month of Madness* (New York: Free Press, 2012).

Clare, Horatio, *Heavy Light* (London: Chatto & Windus, 2021).

Clarke, Roger, *A Natural History of Ghosts: 500 Years of Hunting for Proof* (London: Particular Books, 2012).

De Clérambault, Gaëtan Gatian, *L'Érotomanie* (Paris: Les Empêchers de penser en rond/Le Seuil, 2002).

Dodu, Gaston, 'La Folie de Charles VI', *Revue Historique*, 150(2), 1925, pp. 161–88.

Draaisma, Douwe, *Disturbances of the Mind*, trans. Barbara Fasting (Cambridge: Cambridge University Press, 2009).

Foucault, Michel, *Madness and Civilisation. The Birth of the Asylum* (New York: Vintage Books, 1988, first published 1961).

Grosz, Stephen, *The Examined Life: How We Lose and Find Ourselves* (London: Chatto & Windus, 2013).

Holland, Joanne, 'Narrating Margaret Nicholson: A Character Study in Fact and Fiction' (Ph.D. thesis, Montreal: McGill University, 2008).

Jay, Mike, 'The Art of Mind Control', *Raw Vision*, 59, 2007, mikejay.net/the-art-of-mind-control (accessed 21 January 2022).

Jay, Mike, *James Tilly Matthews and the Air Loom*, revised edn (London: Transworld, 2003).

Jordan-Smith, Paul, *Bibliographia Burtoniana: A Study of Robert Burton's Anatomy of Melancholy* (Redwood City, CA: Stanford University Press, 1931).

Laing, R. D., *The Divided Self* (London: Penguin, 2010, first published 1960).

Lund, Mary Ann, *A User's Guide to Melancholy* (Cambridge: Cambridge University Press, 2001).

MacDonald, Michael, 'The Fearefull Estate of Francis Spira: Narrative, Identity and Emotion in Early Modern England', *Journal of British Studies*, 31(1), 1992, pp. 32–61.

Mantel, Hilary, *A Place of Greater Safety* (London: Viking, 2006).

Murat, Laure, *The Man Who Thought He Was Napoleon*, trans. Dekke Dusinberre (Chicago: University of Chicago Press, 2014).

Nicholson, Margaret, 'Authentic memoirs of the life of Margaret
    Nicholson, who attempted to stab His Most Gracious Majesty
    with a knife…on Wednesday, Aug. 2, 1786', *The Women's Print
    History Project*, 2019.
Nochimson, Richard, 'Studies in the Life of Robert Burton', *The
    Yearbook of English Studies*, 4, 1974, pp. 85–111.
Porter, Roy, introduction to *Illustrations of Madness* (London:
    Routledge, 1988).
Rokeach, Milton, *The Three Christs of Ypsilanti* (New York: New
    York Review of Books, 2011, first published 1964).
Scull, Andrew, *Madness in Civilisation: A Cultural History of Insanity
    from the Bible to Freud, from the Madhouse to Modern Medicine*
    (London: Thames & Hudson, 2015).
Summerscale, Kate, *The Haunting of Alma Fielding: A True Ghost
    Story* (London: Bloomsbury, 2020).
BBC Radio 4, *A History of Delusions*, 2018, and *The Glass Delusion*,
    2015, both available via BBC Sounds. Contributors include
    Adam Phillips, Mary Ann Lund, Andrew Scull, Edward
    Shorter, Dr Emmanuelle Peters and Richard Bentall.

ARCHIVES

The nineteenth-century medical records of major asylums are
    held in the Département de la Seine for the Greater Paris area.
Bethlem Royal Hospital online archives, for admission and dis-
    charge records, hearings evidence, particularly Minutes of
    Bethlem Governors' Subcommittee. Also the 1815 *House of
    Commons Committee Report on Madhouses,* and John Haslam's
    letter of 1818.

Christ Church College and the Ashmole Collection at the
Bodleian Library, Oxford, for all archive related to Robert
Burton and Simon Forman.

The online Wellcome Collection and Catalogue for countless
digitised works and out-of-print classical treatises, and including Nathanial Bacon's *Fearefull* history of Francis Spira in its
many editions.

Parish birth, marriage and death records for James Till(e)y
Matthews and Margaret Nicholson accessed via ancestry.
com and findmypast.com searches.

The online British Newspaper Archive for eighteenth- and nineteenth-century newspapers.

Anon., 'The Plot Investigated', *Annual Register*, 1786, pp. 233–34.

Newspapers carrying the story of Margaret Nicholson's attack
on King George III in 1786: *Scots Magazine*, 1 August 1786
(BL); *Belfast Evening Post*, 10 August 1786; *Hereford Journal*, 10
August 1786; *Kentish Gazette*, 8 August 1786; *Derby Mercury*,
3/10 August 1786; *Chelmsford Chronicle*, 11 August 1786. All
accessed through the online British Newspaper Archive.

Notes

PREFACE

1   Piccolomini, Enea Silvio (Papa Pio II), *I Commentarii*, I, ed. L. Totaro (Milan: Adelphi Edizioni, 1984), p. 1056.

INTRODUCTION

1   See Eaton, W. W., Romanoski, A., et. al., 'Screening for psychosis in the general population', *Journal of Nervous and Mental Disease*, 179(11), 1991, pp. 689–93. This epidemiological point referenced originally by Daniel Freeman in *A History of Delusions*, BBC Radio 4, tx. 2018.
2   Mary Ann Lund, Andrew Scull and Adam Phillips interviewed for *A History of Delusions*, BBC Radio 4.

CHAPTER I

1   Capgras, J., and Reboul-Lachaux, J., 'L'Illusion des "sosies" dans un délire systématisé chronique', *Bulletin de la Société Clinique de Médecine Mentale*, 11, 1923, pp. 6–16. My translation.
2   Ibid. Translation used herein is by Willis, H. D., Whitley, J., and Luauté, J.-P., as it appears in 'Delusional Misidentification: The three original papers on the Capgras, Frégoli and intermetamorphosis

delusions', *History of Psychiatry*, 5(17), 1994, pp. 117–46. Capgras and Reboul-Lachaux also published 'L'Illusion des "sosies"', in *Annales Médico-Psychologiques*, 81(13), 1923.

3   'Der Doppelgänger', *Psychoanalytic contributions to Myth research. Collected studies 1912–1914* (Leipzig and Vienna: Internationaler Psychoanalytischer Verlag, 1919), pp. 267–354.

4   'The Uncanny', 1919, *The Standard Edition of the Complete Psychological Works of Sigmund Freud* (SEXVII), trans. James Strachey and Anna Freud (London: Hogarth Press, 1925).

5   Capgras and Reboul-Lachaux, 'L'Illusion des "sosies"'.

6   Sérieux, Paul, and Capgras, Joseph, *Les Folies raisonnantes: Le Délire d'interprétation* (Paris: Félix Alcan, 1909).

7   See Postel, J., and Allen, D. F., 'Joseph Capgras 1873–1950', *Psychopathology*, 27(3–5), 1994, pp. 121–2; Capgras, J., 'Crimes et Délires Passionnels', *Annales Médico-Psychologiques*, 12(1), 1927, pp. 32–47.

8   Crocq, L., 'La psychiatrie de la Première Guerre mondiale. Tableaux cliniques, options pathogéniques, doctrines thérapeutiques', *Annales Médico-Psychologiques*, 163(3–4), 2005, pp. 269–89. See also Gruselle, Günther, 'Les Troubles Psychiques de la Grande Guerre', conference paper for *Memorial de Verdun: Le Service de Santé dans la Grande Guerre* (Brioude: Éditions Italiques, 2009); Derrien, Marie, 'La tête en capilotade. Les soldats de la Grande Guerre intérnes dans les hopitaux psychiatriques français de 1918 aux années 1980' (Ph.D. thesis, Lyon: L'Université Lumière, 2015).

9   See Warner, Marina, *Fantastic Metamorphoses, Other Worlds: Ways of Telling the Self* (Oxford: Oxford University Press, 2004), pp. 27–8.

10 *L'Annuaire statistique de la France*, vol. xvii, 1897, p. 7.

11 *Illustrated London News*, 6 November 1920, p. 729. John Ptak found a vintage print copy of the edition containing the story in his bookshop and brought it to public attention in 2014.

12 Capgras and Reboul-Lachaux, 'L'Illusion des "sosies"'.

13 In a *Bulletin de la Société Clinique de Médecine Mentale*, 12, 1924, pp. 210–17.

14 Capgras and Reboul-Lachaux, 'L'Illusion des "sosies"', trans. in 'Delusional Misidentification', *History of Psychiatry*, 5(17), 1994, p. 128.

15 Kahlbaum, K. L., 'Die Sinnesdelirien', *Allgemeine Zeitschrift für Psychiatrie*, 23, 1866, pp. 56–78.

16 See Berson, R. J., 'Capgras' syndrome', *American Journal of Psychiatry*, 140, 1983, pp. 969–78, for other cases.

17 See Draaisma, Douwe, *Disturbances of the Mind*, trans. Barbara Fasting (English edn, Cambridge: Cambridge University Press, 2009), pp. 314–15 for these cases; Beers, Clifford, *A Mind that Found Itself, an Autobiography* (New York: Longmans, Green, and Co., 1910), p. 25.

18 Postel, J., and Allen, D. F., conference paper 'The Delusionsal Misidentification Syndromes: Joseph Capgras (1873–1950)', *Psychopathology*, 27(3–5), 1994, pp. 121–2.

19 Thébaud, François, 'La Grande Guerre: La triomphe de la division sexuelle', *Histoires des femmes en Occident*, eds Georges Duby and Michelle Perrot (Paris: Plon, 1992), pp. 31–74.

CHAPTER II

1 *La Justice*, Paris et Départements, Monday 10 February 1890. Translated from French by Kate McCrickard.

2 *London Gazette* as quoted in *Staffordshire Advertiser*, 12 November 1796.

3 Haslam, John, *Illustrations of Madness* (London: G. Hayden, 1810), The Wellcome Collection, pp. 52–55.

4 Haslam, *Illustrations of Madness*, pp. 19–21.

5 *To-day, the popular illustrated magazine*, vol. I, 5 April (Maclean Stoddard and Co., 1873), pp. 428–9.

6 See Ellenberger, Henri, *Discovery of the Unconscious: The History and Evolution of Dynamic Psychiatry* (New York: Basic Books, 1970), p. 64.

7 Clarke, Arthur C., first published in a letter to *Science* magazine, 19 January 1968.

8 *The Moral and Political Works of Thomas Hobbes of Malmesbury*, Part I (London: 1750), p. 128.

9 Esquirol, Jean-Étienne, *Des maladies mentales considérées sous leur rapport medical, hygiénique et médico-légal* (Paris: J. B. Baillière, 1838).

10 Pinel, Philippe, *Traité medico philosophique sur l'aliénation mentale ou la manie* (Paris: Richard, Caille et Ravier, Year IX [1800]).

11 Gale, Colin, *Presumed Curable: An Illustrated Casebook of Victorian Psychiatric Patients in Bethlem Hospital* (Petersfield: Wrightson Biomedical, 2003).

12 Maher, B., 'Delusional thinking and Perceptual disorder', *Journal of Individual Psychology*, 30, 1974, pp. 98–113.

CHAPTER III

1   Unless otherwise specified, I am referring throughout to the Everyman's Library edition of *The Anatomy of Melancholy* (London: J. M. Dent, 1968; first published 1932). The text, edited by Holbrook Jackson, follows the sixth edition, published posthumously in 1651 but accompanied by Burton's last instructions, collated with the fifth edition of 1638, the last in Burton's lifetime. Burton's Latin phrases have been translated and some spelling modernised for clarity.

2   Burton, Robert, 'Democritus Junior to the Reader', *The Anatomy of Melancholy*, vol. 1 (London: Everyman's Library, J. M. Dent, 1968; first published 1932).

3   Wood, Anthony à, *Athenae Oxonienses*, II, col. 653 (London: Thomas Bennet, 1691).

4   'Account of the Author' in Burton, Robert, *The Anatomy of Melancholy* (London: William Tegg, 1863), p. x.

5   Aubrey, John, *Brief Lives*, ed. Andrew Clark (Oxford: Clarendon Press, 1898), vol. 1, p. 130.

6   Wood, *Athenae Oxonienses*, II, col. 653.

7   Burton, 'Democritus Junior to the Reader', *The Anatomy of Melancholy*, vol. 1, p. 18.

8   Ibid., pp. 18–19.

9   Wood, *Athenae Oxonienses*, II, col. 653.

10 Hearne, Thomas, *Remarks and Collections of Thomas Hearne*, XI, 1731–1735 (Oxford: Clarendon Press, 1921), p. 299.

11 Burton, 'Causes of Melancholy: Fears of the Future', *The Anatomy of Melancholy*, vol. 1, p. 366.

12 Ibid., 'Causes of Melancholy: Stars', p. 206.

13 Ibid., 'Causes of Melancholy: Fears of the Future', pp. 364–5.

14 Bodleian Library, 4oR.9Art. See O'Connell, Michael, *Robert Burton* (Boston, MA: Twayne Publishers, 1986), p. 2; Bamborough, J. B., *Robert Burton's Astrological Notebook*, vol. XXXII, issue 127, August 1981; Kiessling, N., 'Two Notes on Robert Burton's Annotations', *Review of English Studies*, New Series, 36(143), 1985.

15 *Oxford Council Acts* 1583–1626, ed. Revd H. E. Salter (Oxford: Clarendon Press for Oxford Historical Society, 1928), pp. 153–4; *Calendar of State Papers Domestic,* James I, 1603–10, p. 35.

16 Burton, William, *Descriptions of Leicestershire*, W. Whittingham, published by subscription, 1777, University of Leicester Special Collections.

17 Gowland, Angus, *The Worlds of Renaissance Melancholy: Robert Burton in Context* (Cambridge: Cambridge University Press, 2006), p. 5.

18 Burton, *The Anatomy of Melancholy*, vol. 1, pp. 39–46.

19 Galen, *On The Affected Parts*, trans. Rudolph E. Siegel (Karger: New York, 1976).

20 Arateus the Capadoccian, *The Extant Works*, ed. and trans. Francis Adams (London: The Sydenham Society, 1856).

21 Quoted in Zilboorg, G., and Henry, G. W., *A History of Medical Psychology* (New York: W. W. Norton and Company Inc., 1941), p. 77.

22 Burton, 'Love-Melancholy', *The Anatomy of Melancholy*, vol. 3, p. 7.

23 Oxford, Bodleian Library, MS Ashmole 226, fol.125r. Simon Forman left his casebooks to his astrological protégé Richard Napier and they became part of the Ashmole Collection. I have also accessed the invaluable digital archive of the Forman casebooks, the work of a team of scholars at University of Cambridge who have transcribed and photographed the paper archive: Kassell, Lauren, et al., *The casebooks of Simon Forman and Richard Napier 1596–1634, a digital edition.*

24 Traister, B. H., 'New Evidence about Burton's Melancholy?' *Renaissance Quarterly*, 29(1), 1976, p. 69. Traister cites a letter Burton wrote from Oxford to his brother William, who was living in London in 1605.

25 See Kassell, L., 'How to read Simon Forman's Case Books; Medicine, Astrology and Gender in Elizabethan London', The Society for the

Social History of Medicine Prize Essay, 1999. Kassell cites William Lilly's *Christian Astrology* (London: 1647).

26 Oxford, Bodleian Libraries, Ashmole MS 226, fol. 125.

27 Ibid., fol. 131v.

28 Ibid., MS 226, fol. 156.

29 Ibid., MS 226, fol. 187v.

30 Ibid., MS 226, fol. 231.

31 Burton, 'Progress of Melancholy', *The Anatomy of Melancholy*, vol. 1, pp. 406–7.

32 Homer, *The Iliad*, trans. R. Lattimore (Chicago: University of Chicago Press, 1951, 2011).

33 Apollonius of Rhodes, *The Voyage of Argo*, Book 4, trans. E. V. Rieu (London: Penguin, 1959), 817ff.

34 Nonnus, 'Dionysiaca', Loeb Classical Library, vol. II, trans. W. H. D. Rouse (Cambridge, MA: Harvard University Press, 1940), 113ff.

35 Aeschylus, *The Complete Greek Tragedies*, trans. H. W. Smyth, Loeb Classical Library, ed. R. Lattimore, 1985ff (Chicago: University of Chicago Press, 1953).

36 Burton, 'Democritus to the Reader', *The Anatomy of Melancholy*, vol. 1, p. 18. The second sentence in the passage, 'Preferment I could not get...Diogenes to his tubbe', is from earlier editions and was replaced in the 1628 edition and thereafter.

37 Lilly, William, *Portrait of his Life and Times* (London: 1715), pp. 43–4, 168.

CHAPTER IV

1 Bacon, Nathaniel, *A Relation of the Fearefull Estate of the Francis Spira, in the yeare 1548* (first published London: Printed by I. L. for Phil. Stephens and Christoph. Meredith, St Paul's Churchyard, 1638), p. 14. The 1672 edition is referred to throughout, printed by Ratcliff, T., and Thompson, N., for Thomas, E., at the Adam and Eve in Little Britain.

2 Burton, 'Religious Melancholy', *The Anatomy of Melancholy*, vol. 3, p. 407. Burton mistakes 1545 for 1548, probably because he had copied

the mistake made in a translation of a French version which had been translated back into English: *Admirable and Memorable Histories containing the wonders of our time. Collected into French out of the best authors. I. Goulart. And out of French into English by Ed. Grimeston* (London, 1607). Burton asserts wrongly about Spira in the *Anatomy* that 'Springer, a lawyer, hath written his life', probably copying another mistake made in the course of back-and-forth translation. 'Springer' was probably a corruption of 'Skrymgeour' the Scot, who had his account of Spira included in the Curione collection of 1550. See 'Lamb, Burton and Francis Speira', *Notes and Queries*, 11(3), 1911, pp. 61–2.

3   Bacon, *A Relation of the Fearefull Estate*, p. 23.

4   See Overell, M. A., 'The Exploitation of Francesco Spiera', *The Sixteenth Century Journal*, 26(3), 1995.

5   Gribaldi, Matteo, *Historia de Quodam (F. Spira), quem hostes Evangelii in Italia coegerunt abicere agnitam veritatem* (1549) was probably the first published account. It was also published in England as *A Notable and Marvellous Epistle*, trans. Edward Aglionby (Worcester: 1550), and included in the collection of accounts of Spira's death, probably edited by Caelio Secundo Curione and published in Basel in 1550 as *Francisci Spierae qui quod susceptam semel Evangelicae Veritatis professionem abnegasset*, pp. 33–56, English trans. Anne Jacobsen Schutte, *Vergerio*, p. 241; Philippe Arles, *The Hour of Our Death*, trans. Helen Weaver (London: Peregrine, 1987), pp. 303–4; also Gribaldi, Matteo, *A notable and marueilous epistle of the famous doctour, Matthewe Gribalde, Professor of the lawe, in the Vniuersitie of Padua: co[n]cernyng the terrible iudgemente of God, vpon hym that for feare of men, denieth Christ and the knowne veritie: with a preface of Doctor Caluine* (London: Imprinted by Henry Denham [and J. Kingston?] for William Norton, 1570), pp. 2, 11, 12.

6   For Martyr's time in England, see McNair, Phillip, 'Peter Martyr in England' in *Peter Martyr Vermigli*, ed. Joseph C. McLelland (Waterloo, Ontario: Sir Wilfred Laurier University Press, 1980), pp. 85–105. Also Anderson, M., 'Rhetoric and Reality: Peter Martyr and the English Reformation', *The Sixteenth Century Journal*, 19(3), 1988, p. 453. The lectures were published: Pietro Vermigli, *In Epistolam S. Pauli Apostoli ad Romanos commentarij doctissimi* (Basel: P. Perna, 1558), and

then translated into English by Sir Henry Billingsly, *Most Learned and Fruitfull Commentaries upon the Epistle of S. Paul to the Romanes* (London: John Daye, 1568).

7 Seaver, Paul S., *Wallington's World* (Stanford, CA: Stanford University Press, 1985), p. 202.

8 Letter to Harding as reported in the *Epistle of the Ladye Jane* (John Day, 1554). See Overell, Muriel Anne, *Italian Reform and English Reformations c.1535–1585* (first published Aldershot: Ashgate, 2008; London: Routledge, 2016), p. 146.

9 Bunyan, John, *Grace Abounding for the Chief of Sinners and The Pilgrim's Progress*, ed. Roger Sharrock (London, New York: Oxford University Press, 1966), p. 51.

10 See Overell, 'The Exploitation of Francesco Spiera'.

11 Interrogation of Spira, May and June 1548; De Leva, *Eretici*, pp. 32–9, as cited in Camponetto, Salvatore, *La Riforma Protestante nell'Italia del Cincequento* (Turin: Claudiana, 1992), p. 64. See Overell, 'The Exploitation of Francesco Spira' for all references to the Interrogation.

12 De Leva, *Eritici*, p. 39; Schutte, *Vergerio*, pp. 239–40. Others also bear witness to the idea that there may have been money-grabbing, especially Henry Scrymgeour whose account is abbreviated in Simon Goulart's *Admirable and Memorable Histories* (London: 1607) pp. 187–96, where Spira is 'a wise and very rich man', p. 187.

13 Burton, 'Religious Melancholy', *The Anatomy of Melancholy*, vol. 3, p. 407.

14 Burton, 'Causes of Melancholy: Fears of the Future', *The Anatomy of Melancholy*, vol. 1, pp. 365–6.

CHAPTER V

1 His father King Charles V lists crystal in his inventory of his own possessions and debts towards the end of his reign. See Miskimin, H. A., 'The Last Act of Charles V: The Background of the Revolts of 1382', *Speculum*, 38(3), 1963, pp. 433–42.

2 Pius Secundus Pontifex Maximus, *The Commentaries of Pius II*, Book VI, trans. Florence Aiden Gragg, *Smith College Studies in History*, 35, 1951, pp. 413–816, 425.

3 *La Chronique du Religieux De Saint-Denys, Contenant Le Règne de Charles VI, de 1380–1422*, published in Latin for the first time and translated by M. L. Ballague (Paris: Imprimerie de Crapelet, 1844).

4 *Les Chroniques de Sire Jean Froissart*, Book IV, ed. J. A. C. Buchon (Paris: Au Bureau du Panthéon Littéraire, 1852).

5 Dodu, G., 'La Folie de Charles VI', *Revue Historique*, 150(2), 1925, pp. 161–88.

6 See account in Pope Pius II, *Commentarii rerum memorabilium quae temporibus suis contigerunt* (Rome: Dominici Basae, 1584), p. 164.

7 Cited in Hainsworth, G., 'La Source du "Licenciado Vidriera"', *Bulletin Hispanique*, XXXII, 1930, pp. 71–2.

8 Howell, James, *Familiar Letters or Epistolae Ho-Elianae* (London: J. M. Dent, 1903), p. 62.

9 Speak, G., '*El Licenciado Vidriera* and the Glass Men of Early Modern Europe', *Modern Language Review*, 85(4), 1990, pp. 850–65.

10 See Ford, Jeremiah G. M., and Lancing, Ruth, *Cervantes, a Tentative Bibliography* (Cambridge, MA: Harvard University Press, 1931), as cited by Engstrom, A. G., 'The Man Who Thought Himself Made of Glass, and Certain Related Images', *Studies in Philology*, 67(3), 1970, pp. 390–405.

11 Alfonso Ponce de Santa Cruz, physician to Philip II Spain, in his book on melancholy, written *c*.1614. This royal is also described by Santa Cruz's friend André du Laurens, the chief physician to Henri IV of France in *Discourses of Melancholike Diseases; of Rheumes, and of Old Age*, trans. Richard Surphlet (Shakespeare Association Facsimiles no. 15, 1938). See Amerzúa y Mayo, A. G., *Cervantes, Creador de la novela corta Espanola* (Madrid: C.S.I.C, 1956–58), p. 161.

12 Du Laurens, André, *A Discourse of the Preservation of the Sight: of Melancholike diseases; of Rheumes, and of Old age*, trans. Richard Surphlet (London: Imprinted by Felix Kingston for Ralph Iacson, St Paul's Churchyard, 1599), pp. 101–3.

13 For comprehensive examples, see Engstrom, 'The Man Who Thought Himself Made of Glass'.

14 Court chronicler José Pellicer de Tovar in his news notice for 16 April 1641: see Pellicer de Tovar, José, *Avisos Históricos 1640–1641*, held at National Library Spain, Madrid.

15 Tomkin, Thomas, *Lingua* (London: 1622) first published in *A Select Collection of Old English Plays*, ed. W. Carew Hazlitt, 4th edn (London: Robert Dodsley, 1744; repr. London: Reeves and Turner, 1874–76), p. 350.

16 Huygen, Constantijn, "*'t Costelick Mall*' (1622), lines 103–8, in *De Gedichten van Constantijn Huygens*, ed. J. A. Worp (Groningen: J. B. Wolters, 1892–99).

17 Descartes, René, *A Discourse on Method: Meditations and Principles*, trans. John Veitch, 1901 (London: J. M. Dent, 1969).

18 Dupré, Ernest, Achalme, Pierre Jean, and Bourget, Paul, *Pathologie de l'imagination et de l'émotivité* (Paris: Payot, 1925).

19 Dupré, E., 'La Folie de Charles VI, Roi de France', *Revue des Deux Mondes*, 60(4), 1910, pp. 835–66.

20 Baring-Gould, Sabine, *The Book of Werewolves* (London: Smith, Elder & Co., 1865), pp. 145–6, as cited in Engstrom, 'The Man Who Thought Himself Made of Glass'.

21 *Le Licencié Vidriera: Nouvelle traduite en Français*, ed. R. Foulché-Delbosq (Paris: Librairie H. Walter, 1892), p. 32.

22 James, William, *The Principles of Psychology* (New York: Henry Holt & Co., 1890) II, 114f; see also Beveridge, A., 'Voices of the mad: patients' letters from the Royal Edinburgh Asylum, 1873–1908', *Psychological Medicine*, 27(4), 1997, pp. 899–908.

23 Pottier, C., 'Perturbations de l'image corporelle dans un cas de psychose hallucinatoire chronique', *Annales Médico-Psychologiques* CII (1944), 1870–90, as cited in Engstrom, 'The Man Who Thought Himself Made of Glass'.

24 Anon., 'Turkey: the Impatient Builder', *Time* magazine, LXXI, No. 5, 3 February 1958, p. 20.

25 Heller, Erich, *The Ironic German, a Study of Thomas Mann* (London: Secker & Warburg, 1958), p. 286.

CHAPTER VI

1 *London Gazette Extraordinary* within the *Scots Magazine* of August 1786.
2 Ibid.
3 *Hereford Journal*, 10 August 1786.
4 Anon., *Authentic Memoirs of the Life of Margaret Nicholson* (London: James Ridgeway, 1786); followed by Anon., *The Plot Investigated* (London: E. Macklew, 1786).
5 'The Examination of Margaret Nicholson', 42/9/455-456, Public Record Office, Kew.
6 *Hereford Journal*, 17 August 1786.
7 *Chelmsford Chronicle*, 11 August 1786.
8 *Ipswich Journal*, 26 August 1786.
9 *Hereford Journal*, 17 August 1786.
10 *Belfast Evening Post, Dublin Evening Post*, 28 August 1786; *Caledonian Mercury*, 30 August 1786.
11 The Parish Registers of Stokesley, Co. York, 1571–1750, ed. John Hawell (Leeds: Yorkshire Parish Register Society, 1901), p. 227.
12 Knollys, William, in a letter to the Countess of Banbury, 30 August 1786, National Archives, ref: 21M69.
13 'The Examination of George Nicholson', HO 42/9/457, Public Record Office, Kew.
14 Old Bailey proceedings, 30 May 1781.
15 *Scots Magazine*, 12 August 1786.
16 *The Cambridge History of English and American Literature*, vol. XIV, eds A. W. Ward and A. R. Waller (Cambridge: Cambridge University Press, 1907–21).
17 *Sophie in London, 1786, Being the diary of Sophie von La Roche*, trans. Clare Williams (London: Jonathan Cape, 1933), pp. 166–71.
18 Cambray, Jacques, *De Londres et de ses environs* (Amsterdam: 1788), ii, pp. 12–13.
19 Anon., *Sketches in Bedlam* (London: Sherwood, 1823), pp. 253–8.

CHAPTER VII

1   See Pinel, Philippe, 'Tableau general des fous de Bicêtre', in Weiner, Dora, *Comprendre et soigne: Philippe Pinel (1745–1826), la médecine de l'esprit* (Paris: Fayard, 1999), p. 143, as cited in Murat, Laure, *The Man Who Thought He Was Napoleon* (Chicago: University of Chicago Press, 2014).

2   For all references to clockmaker case see: Pinel, Philippe, *Treatise on Insanity*, trans. D. D. Davis from the French 1801 (London: Caddel and Davies, 1806), pp. 69–72, as cited in Murat, *The Man Who Thought He Was Napoleon*.

3   Pinel, Philippe, *Traité medico-philosophique sur l'aliénation mentale ou la manie* (Paris: Richard, Caille et Ravier, 1800), pp. 50–51; cf. Pinel, *Treatise on Insanity*, p. 52.

4   According to the actor François-Joseph Talma, *Memoire de J-F Talma*, 1849 (Montreal: Joyeux Roger, 2006), p. 219.

5   Pinel, Philippe, letter to his brother, 8 December 1778, *Lettres de Pinel*, p. 37. See Murat, *The Man Who Thought He Was Napoleon*.

6   Pinel, *Traité*, pp. 66–70.

7   Goldstein, Jan, *Console and Classify: The French Psychiatric Profession in the Nineteenth Century* (Cambridge: Cambridge University Press, 1990), pp. 78–9.

8   'Registre d'observations médicales hommes et femmes', Charenton, 1827. See Murat, *The Man Who Thought He Was Napoleon*, for details on all these cases.

9   Borden, Mary, *The Forbidden Zone* (London: William Heinemann, 1929), p. 159.

CHAPTER VIII

1   'Registre d'observations médicales hommes et femmes', Charenton, 1831, ADVDM, 4X699, fol. 77; see Murat, *The Man Who Thought He Was Napoleon*.

2   See Murat, *The Man Who Thought He Was Napoleon*. Murat cites Alphonse Esquiros quoting Dr Voisin in his survey of Paris, *Paris, ou*

les sciences, les institutions et les moeurs au XIXe siècle (Paris: Au comptoir des Imprimeurs Unis, 1847), vol. 2, p. 118.

3 Nineteenth-century medical records in the Département de la Seine for the major asylums of the Greater Paris Area.

4 Stompe, T., et al., interviewed more than 1,000 people with delusions looking into cultural influences on delusions for 'The pathoplastic effect of culture on psychotic symptoms in schizophrenia', *World Cultural Psychiatry Research Review*, 1(3–4), 2006, pp. 157–63.

5 Patient admitted 16 June 1852; transferred 20 July 1856. See 'Registre d'observations médicales hommes et femmes', La Salpêtrière Hospital (1851–4), ref: 5th division, 2nd section, 6R24, fol. 108.

6 Hobbes, Thomas, *Leviathan* VIII, 1651.

7 Fabre, Antoine, *Bibliothèque du médecin-practicien*, vol. 9 (Paris: J. B. Baillière, 1849), pp. 494–5.

8 Esquirol, Jean-Étienne, *Mental Maladies* (1820), trans. E. K. Hunt (Philadelphia: Lea and Blanchard, 1845), p. 210.

9 Ibid., p. 44.

10 Account contained within a review inside the *Analectic Magazine* of John Reid's 'Essays on hypochondriacal and other nervous affectations', vol. X, July–December 1817, M. Thomas, Philadelphia.

11 Sade, *Journal inédit*, p. 116. Quoted in Murat, *The Man Who Thought He Was Napoleon*, p. 92.

12 Fabre, *Bibliothèque du médecin-praticien*, vol. 9, p. 496.

13 Rokeach, Milton, *The Three Christs of Ypsilanti* (New York: Knopf, 1964; afterword added 1981; repr. *New York Review of Books*, 2011).

14 Beccaria, Cesare, *On Crimes and Punishments* (Livorno: Marco Coltellini, 1764). Commentary attributed to Voltaire, to Cesare, *On Crimes*, translated from French, fourth English Language edn (London: F. Newbery, 1775).

15 *Harper's Magazine*, December 1954; Lindner, Robert, *The Fifty-Minute Hour* (New York: Bantam, 1958), pp. 193–4.

16 See Bell, Vaughan, 'Jesus, Jesus, Jesus', Slate, 26 May 2010, slate.com/technology/2010/05/the-three-christs-of-ypsilanti-what-happens-when-three-men-who-identify-as-jesus-are-forced-to-live-together.html (accessed 21 January 2022).

17 Patient admitted 10 June 1831; discharged 7 September 1831. See 'Registre d'observations médicales hommes et femmes', Charenton, 1831, ref: 4X699, fol. 77.

18 See 'Delusions Associated with Consistent Pattern of Injury' containing interviews with delusional patients by Devinsky, Orrin, NYU Langone Medical Center, New York University School of Medicine, 2009.

19 The 1904 silent film *Maniac Chase*, made by the Edison Manufacturing Company in the US, shows a Napoleon in a cell, wearing the time-honoured bicorn hat and frock coat, fighting with his guards. He escapes, is chased, hides in a barrel and a tree, runs back and eventually does a full circle and ends up back in his cell. There is something about how Napoleon keeps coming back that is well expressed by the circularity of this little film.

CHAPTER IX

1 Cotard, J., 'Du délire hypocondriaque dans une forme grave de la mélancholie anxieuse', *Annales Médico-Psychologiques*, 4, 1880, pp. 168–74. For translation see Berrios, G., and Luque, R., 'Cotard's "On hypochondriacal delusions in a severe form of anxious melancholia"', *History of Psychiatry*, 10(38), 1999, pp. 269–78.

2 Burton, 'Music a Remedy', *The Anatomy of Melancholy*, vol. 2, p. 115.

3 See Förstl, H., and Beats, B., 'Charles Bonnet's Description of Cotard's Delusion and Reduplicative Paramnesia in an Elderly Patient (1788)', *British Journal of Psychiatry*, 160, 1992, p. 417.

4 See Pearn, J., and Gardner-Thorpe, C., 'A Biographical note on Marcel Proust's Professor Cottard', *Journal of Medical Biography*, 11(2), 2003, pp. 103–6, on comparisons between two characters.

5 Proust, Marcel, *In Search of Lost Time, II: Within a Budding Grove*, trans. C. F. Scott Moncrieff and Terence Kilmartin, revised D. J. Enright, 1992 (London: Chatto & Windus, 1981; revision 1992; repr. London: Vintage, 2005), pp. 4–5.

6 Ibid.

7   Ritti, A., *Eloge du Docteur Jules Cotard* (from a letter read at the annual public lecture of the Société Médico-Psychologiques, 30 April 1894) (Paris: Imprimerie de la Cour d'Appel, 1894), pp. 1–10.

8   Jacobi, Maximillian, *On the Construction and Management of Hospitals for the Insane*, trans. John Kitching (London: J. Churchill, 1841).

9   Cotard, Jules, 'Du délire des négations', *Archives de Neurologie*, 11, 1882. For the English translation see Heller-Roazen, Daniel, 'Phantoms: Bodies without organs', *Cabinet*, 25, 2007, cabinetmagazine.org/issues/25/heller-roazen.php (accessed 31 January 2022).

10   Hussey, Andrew, *Paris: The Secret History* (London: Viking, 2006).

11   See Knowlson, James, *Damned to Fame: The Life of Samuel Beckett* (London: Bloomsbury, 1997), p. 208, as cited by Fifield, P., 'Beckett, Cotard's Syndrome and the Narrative Patient', *Journal of Beckett Studies*, 17(1–2), 2008, pp. 169–86.

12   Sacks, Oliver, *The Man Who Mistook His Wife for a Hat* (London: Picador, 1985), p. 2.

13   Berrios G., and Luque R., 'Cotard's Delusion or Syndrome: A Conceptual History', *Comprehensive Psychiatry*, 36(3), 1995, p. 218.

CHAPTER X

1   The collected works of de Clérambault were published by the Presses Universitaires de France as *Oeuvre psychaitrique* (Paris: 1942). The studies focusing on 'Érotomania' also appeared as separate publications: de Clérambault, G., and Brousseau, A., 'Coexistence de deux délires: Persécution et Érotomanie (présentation de malade)' in de Clérambault, G., *L'Érotomanie* (Paris: Les Empêchers de penser en rond/Le Seuil, 2002), pp. 42–64.

2   Hart, Bernard, *The Psychology of Insanity* (Cambridge: Cambridge University Press, 1921).

3   For all de Clérambault's collected papers and lectures on 'Erotomania' see de Clérambault, *L'Érotomanie*.

4   De Clérambault, G., 'Érotomanie pur, Érotomanie associée, Présentation de malade', in G. de Clérambault, *L'Érotomanie*, pp. 79–118.

Index

References to images are in *italics*.